SALAMANDERS OF THE SOUTHEAST

Salamanders

OF THE SOUTHEAST

by Joe Mitchell and Whit Gibbons

The University of Georgia Press   Athens and London

© 2010 by the University of Georgia Press

Athens, Georgia 30602

www.ugapress.org

All rights reserved

Designed by Mindy Basinger Hill

Set in 10/15 Scala

Printed and bound by C&C Offset

The paper in this book meets the guidelines for
permanence and durability of the Committee on
Production Guidelines for Book Longevity of the
Council on Library Resources.

Printed in China

14 13 12 11 10 P 5 4 3 2 1

Library of Congress Cataloging-in-Publication Data

Mitchell, Joseph S. B.

Salamanders of the Southeast / by Joe Mitchell and Whit
Gibbons.

    p.   cm.—(A Wormsloe Foundation nature book)

Includes bibliographical references and index.

ISBN-13: 978-0-8203-3035-8 (pbk. : alk. paper)

ISBN-10: 0-8203-3035-3 (pbk. : alk. paper)

1. Salamanders—Southern States—Identification.
2. Salamanders—Southern States—Classification.
I. Gibbons, J. Whitfield, 1939– II. Title.

QL668.C2M493 2010

597.8'50975—dc22   2009037108

British Library Cataloging-in-Publication Data available

Contents

ALL ABOUT SALAMANDERS

Why Salamanders?   1

Defining the Southeast   5

Biology of Salamanders   5

Salamander Diversity around the World   18

Salamander Taxonomy   22

How to Identify Salamanders   24

SALAMANDER HABITATS IN THE SOUTHEAST

Freshwater Wetland Habitats   32

Terrestrial Habitats   36

SPECIES ACCOUNTS

Organization and Order of Species Accounts   41

Aquatic Salamanders   45

Aquatic Cave Salamanders   84

Stream and Seep Salamanders   95

Seasonal Wetland Salamanders   173

Forest Terrestrial Salamanders   217

PEOPLE AND SALAMANDERS
What Is a Herpetologist? 293
Keeping Salamanders as Pets 298

A CONSERVATION VISION FOR SALAMANDERS 301

**WHAT KINDS OF SALAMANDERS
ARE FOUND IN YOUR STATE?** 306

GLOSSARY 309

FURTHER READING 313

ACKNOWLEDGMENTS 315

CREDITS 317

INDEX OF SCIENTIFIC NAMES 319

INDEX OF COMMON NAMES 321

SALAMANDERS OF THE SOUTHEAST

When you do find that fiery red, traffic light yellow, bright orange, or lichen green salamander, think about the fact that this group of animals has long outlived the dinosaurs.

All salamander larvae are aquatic, have four legs, and develop large gills for oxygen uptake.

All about Salamanders

WHY SALAMANDERS?

Why should someone get excited about small, seldom-seen animals? For one reason, being "out of sight" is no justification for this spectacular group of animals to be "out of mind," especially in the southeastern United States. The remarkably clandestine salamanders are a classic example of hidden biodiversity; they far outnumber all other southeastern terrestrial vertebrate groups in species and often in population sizes. Many people do not know that southeastern salamanders rival the region's other animals in body coloration and patterns, including the dramatic displays of birds, snakes, and butterflies. No fewer than 102 full species are represented in this book, with more destined to be discovered through genetic research. In fact, the salamanders found in the Southeast constitute 17 percent of the world total of 578 known species recognized in early 2009. And they offer many opportunities both to appreciate nature and to understand the importance of even rarely seen animals to the ecology of local communities of plants and animals.

More salamander species live in the tropics than in temperate North America, but more major groups (families and genera) are found in the Southeast than anywhere else. Most people never see salamanders because

The dwarf salamander is one of the few southeastern salamanders that has four toes on the hind feet instead of the usual five.

The lichen-like color and pattern on green salamanders provide camouflage when their background is covered with lichen.

they are so secretive, yet many are major players in aquatic and terrestrial food webs. Some salamander species reach such high population densities that without them their predators might suffer severe impacts, and their prey might rapidly spiral out of control. Unfortunately, the future of an unacceptably high number of salamander species is being placed in jeopardy as their habitat is destroyed or degraded by chemicals and land use alterations.

Even the scientific community has only recently begun to understand how these unique animals communicate with each other, what their relationships are with their environment and with other species, how they can live for so long on only a few meals a year, and why many cannot survive without the full complement of aquatic and terrestrial habitats in the landscape. Indeed, we are only now learning how many species we really have in the Southeast and how they are related to each other genetically. We are learning how they respond to loss of vegetation cover, newly emerging diseases, and input of agricultural chemicals, usually to their detriment. We are also learning how to manage their populations in urban and suburban areas to help ensure that they persist in the Southeast in concert with human activities.

Salamanders may be useful environmental sentinels if only we pay attention to what they are telling us. The first salamanders in the fossil record date back about 230 million years, 40 million years before the appearance of dinosaurs. The salamanders outlived this spectacular group of reptiles by surviving the events that caused the dinosaur extinctions some 65 million years ago. Unfortunately, humans are doing things to the planet that are causing the extinction of salamanders far more rapidly than they can be replaced by evolutionary processes. Amphibians as a whole are declining at an alarming pace. If we do not change our environmental attitudes, humans may very well become the agents of the next great extinction of much of the animal life on Earth—the sixth such event in the planet's history (the fifth being the dinosaur extinctions).

Are amphibians today serving as indicators of the sixth great extinction? Most of the salamander extinctions are occurring in the tropics, but one

Salamanders are related to frogs and toads, and both groups are considered amphibians. All have moist skin and lay unshelled eggs in aquatic or at least moist environments.

This greater siren illustrates the permanently aquatic, elongated salamander body form with only two front legs and gills.

salamander in the Southeast—the Bay Springs salamander (*Plethodon ainsworthi*), a woodland salamander species known from only two individuals caught in Mississippi in 1964—is thought to have become extinct within the last 50 years. Although others may be in jeopardy as a result of small, localized geographic ranges and unregulated environmental impacts, most southeastern salamanders seem to be hanging on, and some are still doing well. What our generation does in the next decade or two will determine how many of our existing salamander species will be around for future generations to study and appreciate.

This book is intended to inspire an appreciation of salamanders through the information it provides and the magnificent photographs taken by numerous herpetologists. Salamanders will not show themselves to you. You must find them by taking a hike or tour to see them in their native habitats or attending a nature class that has a selection of species on hand in captivity. Go out on nature walks in the daytime and turn over logs and rocks (replacing them afterward, of course) to see what lies underneath. Go out with a naturalist who knows how to find salamanders at night or during a rain, when they are most active. We encourage you to learn how to appreciate these amphibians and to value them in the same way that birds and other more visible animals in the Southeast are valued. And when you do find that fiery red, traffic light yellow, bright orange, or lichen green salamander, think about the fact that these animals crawled the earth before the dinosaurs, and that we humans may be edging many of them close to extinction. We hope that you will support efforts to protect and manage habitats all over the Southeast so that our grandchildren and their grandchildren will have opportunities to appreciate the aesthetic beauty of these charismatic, seldom-seen amphibians.

Salamanders, such as this terrestrial Yonahlossee salamander from the southern Appalachian Mountains, are classic examples of hidden biodiversity.

Amphibians are declining at an alarming pace. Ornate chorus frogs, for example, no longer occur in some parts of their range in North America.

Did you know?

The salamanders diverged from the ancestors of both frogs and caecilians some 230 million years ago.

DEFINING THE SOUTHEAST

Regions of the country can be identified from a variety of perspectives, including culture, political boundaries, geological or physiographic regions, and climatic or other environmental features. Clearly, human cultures or politics have little influence on the distribution patterns of salamanders, while the geological history and climate of a region have a great influence. But to continue the practical approach of using states as basic units for describing geographic ranges of species, we have chosen to define the Southeast for the purposes of this book as Alabama, Florida, Georgia, Kentucky, Louisiana, Mississippi, North Carolina, South Carolina, Tennessee, and Virginia. The defined region has an identifiable assemblage of salamanders distinct from those of northern and western states, whose biological character is shaped by different environmental factors and topographic features.

BIOLOGY OF SALAMANDERS

Salamanders are members of the order Caudata ("caudate" means "tail") in the class Amphibia. Other amphibians include frogs and toads (order Anura), which lack tails as adults; and caecilians (order Gymnophiona), a group of legless, wormlike animals found in the tropics. By 2009 scientists had described more than 6,474 species of amphibians worldwide, of which 578 are salamanders; 293 amphibian species live in the United States (107 frogs and 186 salamanders). One hundred and two species of salamanders live in the Southeast as defined in this book. The first fossil salamanders known from North America were found in Arizona and date back to the early Jurassic period, some 200 million years ago. Thus,

Salamanders differ from frogs in having an elongated backbone with 10–60 vertebrae, a long tail, and usually four limbs roughly similar in size.

the modern salamanders in the Southeast are descendants of an ancient group of vertebrates.

The word "amphibian" means "both lives," indicating that most amphibians have both a larval stage that is aquatic and an adult stage that is usually terrestrial. With the exception of some of the salamanders and a few frogs and caecilians, all amphibians remain tied to water just as their ancestors were. Most salamanders either live in water permanently or return to it to lay eggs. The eggs hatch into aquatic larvae that grow for a time, usually weeks or months, before losing their larval characteristics through metamorphosis. Salamanders have thin, permeable skin and must have special behaviors or adaptations to tolerate dry environments.

Salamanders differ from frogs by having an elongated backbone with 10–60 vertebrae, a long tail, and usually four limbs that are roughly similar in size. Frogs have only 5–9 vertebrae, a unique bone called a urostyle in their pelvis, and much enlarged rear legs used for jumping. All salamanders have a tail—usually a long one—a character that distinguishes them from caecilians with their stubby tail and frogs with no tail at all as adults. Salamanders also have a slender, elongated body and a distinct head.

Diversity of Salamanders in the Southeast

The 102 species of salamanders that inhabit the Southeast belong to 17 genera within 7 families. Ten species are members of the family Ambystomatidae, commonly called mole salamanders because they spend much of their lives underground. The mole salamanders have a stocky body, a robust head, and four functional limbs.

The family Amphiumidae includes 3 species of fully aquatic salamanders called amphiumas, 2 of which reach the longest total length of any salamander in the Southeast, a whopping 41 inches. All have an elongated body, two pairs of tiny legs, and internal gills that are not visible on the outside. The only obvious difference among the species of amphiumas is the number of toes on each foot.

The family Cryptobranchidae has only one representative in the Southeast: the hellbender. This large, prehistoric-looking salamander has a wide, flattened head; a robust body with wavy folds of skin along each side; and a paddle-like tail. Other members of this group live in the Ozark Mountains of the Midwest, Japan, and China.

The most diverse group, not only of salamanders but of all vertebrates in the Southeast, is the family Plethodontidae, with 10 genera and 77 species. A major anatomical feature unites all of them: they have no lungs. All their respiration occurs by gills or across the skin. The plethodontids

Two species of newts in the Southeast secrete poison
from skin glands that deters would-be predators from eating them.
This is a red-spotted newt.

Nineteen species of salamanders in the genus
Desmognathus live in rocky streams and adjacent
moist forest habitat in the Southeast.

Thirty-eight species of
terrestrial salamanders in
the genus *Plethodon* are
found in the moist forests
of the Southeast. All lack
lungs and do not have an
aquatic larval stage.

The ambystomatids, commonly called mole salamanders
because they spend much of their lives underground, have
stocky bodies and robust heads. Left to right: mole, marbled,
tiger, spotted, and Mabee's salamanders.

Mudpuppies like this one and its relatives the
waterdogs live in permanent streams in the
Southeast. All are fully aquatic as adults and retain
external gills to breathe in their aquatic habitats.

have diverse body shapes, but all have four limbs, a long tail, a groove that runs from the nostril to the lip (nasolabial groove) on the snout, and movable eyelids.

The Proteidae includes 5 species of mudpuppies and waterdogs, all of which are aquatic throughout life. They are paedomorphic (i.e., the adults retain larval traits) and have external gills, short limbs, and a tail that is flattened side to side.

The 2 members of the Salamandridae in the Southeast are called newts. Unlike all the others, these salamanders have granular skin embedded with poison glands that make them distasteful to predators. Their brightly colored body or belly serves as a warning that they are toxic.

The Sirenidae includes 2 genera and 4 species. All of the sirens are fully aquatic, eel-like in shape, have external gills, and completely lack the rear pair of legs and pelvic girdle.

Most salamanders in the Southeast are too small to bite humans, but some, such as amphiumas and hellbenders, can bite hard and even break the skin. Some, such as the newts, produce toxic secretions in skin glands that would make someone who swallowed one sick. So be careful when picking up a large salamander, and don't swallow any of them.

Activity and Energy

Salamanders, like all amphibians, are ectothermic, or "cold-blooded." Their body temperature approximates that of the surrounding environment because they cannot generate their own body heat from internal chemical reactions as mammals and birds do. Many salamanders are nevertheless active at surprisingly low temperatures. Tiger salamanders, for example, are well known for walking across snow to enter breeding ponds in winter and move about readily under ice cover. Their enzymes function best at low temperatures, and they are inactive in the warm summer months. The ectothermic lifestyle allows amphibians to spend less energy on keeping warm and more on growth and reproduction. Most require only a few meals per year to gain enough energy to survive, grow, and reproduce; and they can survive long periods without food. Salamanders do not hibernate in the usual sense. Some survive cold winter temperatures by seeking out refugia in forest soils or crevices below the frost line and becoming inactive.

Ectotherms are not as active and visible as most birds and some mammals; some species are above ground only a few times during an entire year, and only during night rains. About 90 percent of any woodland salamander (genus *Plethodon*) population remains underground at any given time, while 10 percent is above the ground foraging. Many species have

Did you know?

Glands in the skin of red-spotted newt efts produce secretions that are toxic to many animals, including birds and other would-be predators as well as humans.

highly seasonal activity patterns that are largely regulated by temperature and rainfall. Finding them thus requires knowledge of when and under what conditions they are likely to be active.

The majority of the southeastern salamanders belong to the family Plethodontidae, all of whose members lack lungs. The adults of most species respire through their moist skin. They inhabit highly oxygenated streams or seepage areas

Salamanders store a considerable amount of energy in their tail in the form of fat. If the tail is broken, the salamander can regenerate another.

or, if terrestrial, live in moist forests and remain inactive under surface objects or underground unless the ground is moist and the air humid from rains. The aquatic larvae of plethodontid salamanders and cave-dwelling adults use gills for respiration. All other salamanders have lungs as adults and gills as larvae, but also respire through the skin to at least some degree, and thus must live in humid environments.

Salamanders as Predators

All larval and adult salamanders are carnivores; that is, they eat other animals. Most adults consume a variety of invertebrates such as insects and their larvae, worms, millipedes, centipedes, and snails. Aquatic larvae eat zooplankton, mosquito larvae, and other invertebrates. Some salamanders will kill and eat larvae and adults of other species, and some even cannibalize their own kind. Large tiger salamanders can eat small mice. Amphiumas eat crayfish, tadpoles, and insects. Hellbenders prey heavily on crayfish and occasionally on small fish.

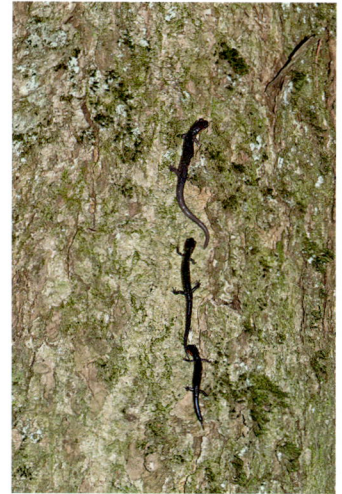

Some salamanders, like these Jordan's red-cheeked salamanders, climb trees and vegetation to obtain prey on warm, moist nights. They are active on the ground surface only when temperatures are moderate and sufficient moisture is present.

Terrestrial red-backed salamanders in Appalachian forests occur in such large numbers that their biomass may exceed that of all the birds and mammals that also occupy their habitat. They are responsible for an enormous amount of the energy that is transferred along the food chain from insects and other species that eat plants to the top carnivores in the ecosystem. Other salamanders can reach high densities in aquatic and terrestrial habitats as well, and all play a crucial role in energy dynamics in aquatic and terrestrial ecosystems. Salamander larvae link the two ecosystems because they transfer the energy they accumulated in the aquatic environment to

Spring salamanders, found in southeastern mountain streams and adjacent terrestrial habitats, are large enough to eat other salamanders, even large ones such as this slimy salamander.

All salamanders are carnivores and obtain energy from a wide variety of invertebrate and small vertebrate animals.

the terrestrial environment when they metamorphose and emerge from wetlands.

Salamanders capture prey in several different ways. They detect potential prey visually or by smell. The size of the prey they can capture and swallow is in part determined by how wide they can open their mouth (gape), which in turn is related to their body length. Salamanders in terrestrial environments, such as mole salamanders, capture their prey by lunging a short distance with the mouth open; on contact they quickly extend the tongue forward and downward so that the back of the tongue contacts the prey and at the same time releases a sticky secretion that adheres to the prey. The tongue—and the insect or worm stuck to it—is quickly withdrawn into the mouth, which closes over it. Large prey are held in place by the teeth that line the upper and lower jaws. The plethodontid salamanders use a similar mechanism for feeding. Although their tongue is not connected in the front as it is in mole salamanders, they can protrude the tongue a short distance. The combination of a short lunge, open mouth, sticky protrusible tongue, and snapping the jaws shut is an effective mechanism for capturing prey.

Feeding in water requires a different strategy. Most salamander larvae and fully aquatic adults essentially vacuum their prey into their mouth. As the salamander makes a brief lunge at the prey, the floor of the mouth drops and the nostrils close, creating a vacuum that literally sucks the prey into the mouth. The mouth closes only partially to allow excess water to escape as it closes without losing the prey. Hellbenders are unique because they can drop one side of their jaw while keeping the other side closed. This allows them to capture prey on either side of the mouth rather than in front of it as most other salamanders do.

Defense Mechanisms

A host of other animals eat salamanders; some predators eat only one or two species exclusively. Large adult mud snakes, for example, eat only sirens and amphiumas. The large population size and biomass of many salamander populations offers predators an ample supply of food. In response, salamanders have developed a variety of ways to avoid becoming a meal.

Most commonly, both aquatic and terrestrial salamanders simply flee. Watch a slimy salamander quickly disappear down a hole as soon as you turn over the log under which it was hiding, for example; or try to catch a tiger salamander larva under the ice with a dip net and watch it avoid your efforts with twists and turns as it swims quickly away. Once caught, however, salamanders have other tricks to make the predator let them go. Large salamanders such as amphiumas and hellbenders can bite hard and rake their teeth across the predator's skin. Small salamanders can bite, too, but it is effective only with small predators. Many terrestrial salamanders that live under logs and rocks freeze in place when uncovered, taking advantage of visual predators' tactic of searching for a moving prey item; they tend to overlook a camouflaged, unmoving salamander. Small salamanders also rely on being distasteful or toxic. Newts secrete a powerful toxin from glands in their skin. This is why they can live in water with fish where species lacking toxins cannot survive.

Unlike the shelled eggs of reptiles, salamander eggs are protected only by several layers of viscous jelly, which makes them easy targets for predators such as predaceous invertebrates. Eastern newts are well known to eat eggs of tiger and spotted salamanders. Fish, giant water bugs, water tigers (an aquatic beetle larva), dragonfly nymphs, predaceous diving beetles, and snakes and wading birds are among the predators that eat salamander larvae, especially when the water is low and the larvae are concentrated. Marbled salamander larvae, which hatch early and are large by the time tiger

Salamanders often distract a potential predator away from the head and body by waving or wiggling the tail. This red salamander is coiling its tail and exposing its undersurface.

Amphiumas and sirens are the primary prey of mud snakes in the Southeast. This mud snake is swallowing an amphiuma.

and spotted salamander larvae hatch in the same pond, eat their smaller relatives. The tiger salamander larvae grow faster, however, and a couple of months later the tables turn, and the tiger salamander larvae are able to eat the marbled salamander larvae.

Some salamanders, such as the terrestrial juvenile eft stage of the red-spotted newt, are bright orange or red, warning would-be predators that they are toxic. A few species found in the same area, such as the red salamander, mimic the toxic efts and are also brilliantly colored, and wary predators pass them by. Many of the woodland salamanders in the genus *Plethodon* secrete sticky, slimy mucus from skin glands that has been known to glue a predator's mouth shut. Although the mucus is nontoxic to humans, it is almost impossible to wash off and will remain on the skin for a day or two. Many salamanders escape predation by waving their tail to distract the predator's attention to that part of the body. Once the predator grabs the tail, the salamander can "autotomize" it, and the predator is left holding a wiggling tail while the rest of the salamander escapes. A new tail will grow to replace the one lost.

Behavior

Unlike frogs, which vocalize to attract mates and defend territories, most salamanders lack the ability to make sounds. Sirens and amphiumas on rare occasions produce a low whistle or make clicking sounds; mole salamanders can whistle, click, or squeak; and some plethodontids can make soft squeaks. Little is known about these seldom-heard sounds, including why they are made, but they may be associated with defense or perhaps are a reaction to being captured and restrained.

This mud salamander is coiling its tail over its head to protect itself from a potential predator.

Salamanders communicate primarily with chemicals and touch. Visual communication is limited because salamanders are generally underground, active at night, or in water with low visibility. All species produce pheromones—small-molecule chemical messengers designed to produce a specific response. Male woodland salamanders, for example, mark the boundaries of their small territories with pheromones from glands located posterior to the cloaca. The owner actively defends his territory, which is often limited to a rock or a portion of a small log, against intruders of the same or different species. Intruders

Salamanders communicate largely by chemicals and touch during courtship and when evaluating potential intruders. The nasolabial groove on the front of the snout picks up pheromone molecules and transfers them to the nasal cavity, where they stimulate sensory neurons that send messages to the brain for interpretation.

Salamanders fight for several reasons, including territorial defense, capturing prey, and courting females. These two Black Mountain salamanders are locked in combat.

know from the smell of the pheromone that the rock belongs to another salamander. The nasolabial groove on the front of the snout picks up the pheromone and transfers it to the nasal cavity. Sensory neurons transmit the signal from there to the brain, where it is interpreted. Intruders either avoid the territorial male or engage in a battle in which biting and wrestling determine the winner; usually it is the owner. Chemical and tactile communication also occurs between males and females as described below. Salamanders' chemical communication system is so sensitive that individuals can recognize other individuals, including their own brothers and sisters. Researchers are currently investigating the complex social systems of salamanders and the possibility that they may have short-term memory and cognitive thinking.

Reproduction

Salamanders accomplish reproduction with an amazing variety of behaviors that range from simple external fertilization of eggs to complex and elaborate rituals in which the male entices the female to pick up a packet of his sperm (i.e., a spermatophore) with her cloaca. The eggs are then fertilized internally. Some species lay eggs in water and have an aquatic larval phase; others are terrestrial and completely skip the aquatic phase.

Courtship in the mole salamander group (Ambystomatidae) is simple and fast-paced. Males arrive at wetlands before females and deposit rows of spermatophores. Females arrive days or even weeks later. A male approaches a female and nudges her body and cloaca. He directs her to straddle a spermatophore by walking backward in front of her, and with additional nudging moves her to position her cloaca over the spermatophore so that she can pick up the sperm packet. Fertilization is internal, and the female then lays her eggs along grass blades or stems of other vegetation. Adults of some species provide parental care. They leave the pond, and the eggs mature and hatch into aquatic larvae several days to weeks later. Marbled salamanders use a variation on this theme that allows them to produce young in the absence of water. Males deposit spermatophores on land during fall rains and entice females to pick up the sperm packets. Females deposit their eggs in loose clusters under logs and debris in the dry pond basin and stay with them until winter rains fill the pond with water. The eggs hatch shortly afterward, and the females make their way back to underground retreats in the forest.

Our knowledge of courtship in amphiumas comes from one observation in an outdoor enclosure of a female that glided over a male lying still in shallow water and positioned her cloaca opposite his. A sperm packet was apparently transferred during the contact. Following internal fertilization, female amphiumas lay eggs in pockets in the substrate, and these hatch into aquatic larvae. Mudpuppy courtship behavior is unknown. The male produces a spermatophore that is somehow transferred to the female, and fertilization is internal. Females provide parental care for their eggs.

Although external fertilization is the usual mode of reproduction among frogs, it is unusual among salamanders. Biologists postulate that fertilization in sirens is external but do not know the specifics because it has never been described. Hellbenders also reproduce by external fertilization. A male entices a female to enter a chamber under his rock in the stream,

Egg follicles enlarge prior to fertilization and egg laying. The presence of yolking eggs in the belly of a female is an indication that the egg-laying season is about to begin.

and she lays long strands of eggs on the underside of the rock. He then moves over them and extrudes white, ropy masses of sperm that fertilize the eggs. An additional oddity with hellbenders is that the male, not the female, remains with the eggs and protects them from potential predators. The eggs hatch into larvae with gills.

Courtship behavior in the lungless salamander group (the Plethodontidae) is remarkably uniform among species. The following sequence is typical whether it occurs in water or on land: (1) the male orients to and approaches a female; (2) the male makes head contact with female; (3) the pair engages in a "tail-straddling walk" in which the female places her chin on the base of the male's tail and positions herself over his tail; (4) the male deposits a spermatophore in front of her snout; (5) the pair moves forward and the female lowers her cloaca over the spermatophore and picks up the sperm packet with her cloacal lips. Fertilization is internal. Males in this group have a remarkable way of enticing a female to be receptive that is called "vaccination." During the mating season, the male grows large teeth in the front of his upper jaw, and the mental gland under his lower jaw grows and enlarges. During the elaborate courtship, the male uses his enlarged front teeth to scratch the skin of the female. He then presses his enlarged mental gland against the abraded skin and releases a secretion from it that enters the female's body and presumably makes her receptive for mating. Females then lay their eggs under logs or rocks or in depressions in the soil and guard them from predators until they hatch. Females of some species produce antibacterial, antifungal skin mucus that protects the developing eggs. Eggs of aquatic plethodontid salamanders develop into gilled larvae; all of the larval development of terrestrial salamanders occurs within the egg, and the hatchling looks like a miniature adult.

Newts have a complex courtship ritual that is easily observed in an aquarium. The male rubs his forearm across the female's snout, then presses his cheek against her snout as he lifts his front leg under her chin. He grasps her around the neck in front of her front limbs, fans his tail back and forth, jerks his tail base, and thrashes his body while continuing to rub her snout with his cheek, all the while dragging the female roughly about. After several minutes he dismounts, moves in front of the female, and while moving forward wiggles his body and arches his tail. The female

Reproductive behavior in plethodontid salamanders such as these Jordan's red-cheeked salamanders is tactile, chemical, and visual. The male rubs his chin on the female to transfer pheromones that will make her receptive to mating.

The female arches her tail when the male straddles her and rubs his mental gland on the top of her tail. This "straddle walk" is a common mating behavior in terrestrial salamanders.

Male spotted salamanders migrate from underground terrestrial retreats to ephemeral pools in late winter. They deposit spermatophores (left) on the substrate that females pick up and use to fertilize their eggs internally. Following internal fertilization, female spotted salamanders attach their eggs (right) to twigs and grass stems in shallow water. The eggs will develop for several weeks before hatching.

The jelly surrounding spotted salamander eggs takes up water and swells to form a protective gelatinous mass that may be clear or opaque.

Hatchling spotted salamanders (left) have gills and all four limbs. These aquatic larvae will grow and develop in the water until they metamorphose into juveniles. The metamorphs (right) are small and often have remnant gill buds behind the head. The juveniles will live in the terrestrial environment for another 2–3 years before they mature and return to the pond to mate.

nudges his cloacal region or his tail, and the male deposits a spermatophore on the bottom. He then moves forward a body length and turns sideways to block her path. She positions herself over the spermatophore and picks up the sperm packet with her cloaca. During the entire process the male is often harassed by other males attempting to dislodge him, although they usually are not successful. The eggs are fertilized internally, and the female deposits each egg singly on a leaf blade and folds it over, gluing it shut with mucus from her cloaca. Egg laying may take several days. Newts show no parental care, and the aquatic larvae hatch in 3–4 weeks.

Age and Longevity

People often equate small body size with short lives and large body size with long lives, but salamanders are surprisingly long-lived for their relatively small size. Red-spotted newts, for example, metamorphose from the larval stage to the eft (juvenile) stage several months after egg laying. The terrestrial efts then forage in the forest for 2–7 years before they mature and return to their home pond to reproduce for the first time. Adults often live for another decade after that. Longevity in natural populations is largely unknown, but in a long-term South Carolina study, individually marked mole salamanders were recaptured more than a decade after they were marked.

Most of the available records on longevity are from long-term captives. Small salamanders such as Blue Ridge two-lined, northern dusky, and northern slimy salamanders have been kept in captivity for as long as 5 years. Larger salamanders such as amphiumas and spotted salamanders are known to live more than 25 years. Of all southeastern salamanders, the hellbender is the longest-lived—more than 30 years. Age at maturity for the small salamanders is often 1–2 years but is as long as 5–10 years for the largest salamanders. Some deer, rabbits, and songbirds, in contrast, reach maturity within a year or just a few months of birth. Many salamanders, that is, live much longer than almost all game species. Among other things, this means that salamander populations are very slow to recover when reproductive adults are lost as a result of habitat destruction or alteration, even if the habitat rebounds to its natural state. Think about the former natural places you know that have been changed in some way, such as land where the hardwood trees were removed through timber harvesting. We should not be surprised that salamanders throughout the Southeast are declining in numbers. Specific conservation issues are noted in each of the species accounts and are summarized later in this book.

The age of salamanders can be determined only by marking them as first-year hatchlings or metamorphs, or by taking a slice out of a leg bone and analyzing the number of growth rings under a microscope.

Did you know?

Many salamanders use tail waving to distribute chemicals known as pheromones to attract mates. An uplifted tail can also distract potential predators away from the salamander's head and body.

The Siberian newt (*Salamandrella keyserlingii*) is found as far north as 64° north latitude in the Russian Plain—almost to the Arctic Circle. It can withstand long periods of being completely frozen.

SALAMANDER DIVERSITY AROUND THE WORLD

Most of the world's 578 recognized species of salamanders are found in the Northern Hemisphere; the group is entirely absent from the continents of Antarctica, Australia, and most of Africa. The only tropical salamanders are in the family Plethodontidae, and they occur mostly in the mountains of Central America and the Amazon basin. These ancient amphibians have adaptations that allow them to occupy many types of freshwater and terrestrial habitats. No salamanders anywhere in the world can tolerate salt water; thus no salamander species are found in oceans, salt marshes, or tidal creeks. Species of some salamanders are found on barrier islands of the Atlantic and Gulf coasts, but they are restricted to freshwater habitats or to woodland habitats not regularly flooded by saltwater.

Many salamander species are found only in high mountains, especially in the Appalachians and in Central America. The Siberian newt (*Salamandrella keyserlingii*) occurs as far north as 64° north latitude in the Russian Plain—almost to the Arctic Circle. This amazing salamander has the largest range of any species in the world and is capable of surviving after being frozen in the tundra ice for several years. One such individual excavated from a depth of 33 feet was estimated to be 90 years old! Although the Sibe-

European fire salamanders (*Salamandra salamandra*) from Eastern Europe exhibit phenomenal variation in colors and patterns. These salamanders got their common name when they were observed emerging from logs put on fires.

rian newt is the extreme case, many salamanders can tolerate subfreezing temperatures for varying lengths of time because they have an antifreeze compound called glycogen in their tissues and organs. Salamanders tend to be more active at and more tolerant of cooler temperatures than frogs.

With more than 100 species, 16 genera, and 7 families, the Southeast indisputably has the highest genetic diversity of salamanders in the United States or any other comparably sized region in the world. Three regions in North America outside the Southeast contain high diversities of salamanders as well: the mountains of the Pacific Coast; the Edwards Plateau region in central Texas; and the Ozark and Ouachita mountains in Arkansas, Missouri, and Oklahoma. Two families—Dicamptodontidae with 4 species in the genus *Dicamptodon*, and Rhyacotritonidae with 4 species in the genus *Rhyacotriton*—occur exclusively in the mountains of the West Coast, along with 3 ambystomatids, 16 plethodontids, and 3 newts. Several species of plethodontids, all in the genus *Eurycea*, occur only in the streams and caves of the Edwards Plateau. Seven members of 4 genera (*Ambystoma*, *Eurycea*, *Plethodon*, *Typhlotriton*) occur primarily in the Ozark region. One is the bizarre grotto salamander (*Typhlotriton spaeleus*), which has adapted to life in caves. Adults are white to pinkish white with black

Did you know?

The largest salamander in the world is 4 feet, 9 inches long and lives in Japan. The Chinese giant salamander gets larger, but the biggest ones are believed to be gone as a result of habitat loss and human consumption.

Marbled newts (*Triturus marmoratus*), which occur in France, Portugal, and Spain, are rare and protected. The skin gets a velvet-like texture and becomes water-repellant during the terrestrial stage.

Members of the tropical genus *Oedipina*, such as this endangered *Oedipina poelzi* (left) from Costa Rica, are characterized by an exceptionally long tail. The Mexican axolotyl (*Ambystoma mexicanum*) lives in lakes around Mexico City. (right) These salamanders never metamorphose and reach enormous sizes.

The Mexican mushroom salamander (*Bolitoglossa mexicana*) is a high-elevation salamander found in Central America.

Members of the genus *Thorius* (left) are the smallest salamanders on earth. These tiny salamanders occur on isolated mountaintops in Mexico. The bizarre grotto salamander (*Typhlotriton spaeleus*) of Texas has completely adapted to life in caves (right). Adults are white to pinkish white with black spots for eyes and spindly legs.

spots for eyes and spindly legs. The sirens (family Sirenidae) are confined to midwestern and southeastern North America. All of the members of the Amphiumidae (*Amphiuma*) are found in the Southeast. Members of the Cryptobranchidae occur in North America (hellbenders) and Asia (2 species of giant salamanders are found in China and Japan).

Only one family occurs exclusively outside North America. The Asiatic salamanders in the family Hynobiidae inhabit Asia and most of Russia. Most are less than 4 inches in total length, but one, *Ranodon sibericus*, reaches almost 10 inches. Except for the European olm (*Proteus*, in the family Proteidae), all mudpuppies and waterdogs are confined to North America. The newt family (Salamandridae) has the largest global distribution of any group of salamanders. Several species are found in Europe, Asia, and eastern and western North America. Many of the newts are brightly colored, and most are quite toxic as a result of the poison secreted by their abundant skin glands. Several, such as the European fire salamander (*Salamandra salamandra*), actually squirt their poison from glands at predators. Two European species of newts bear their young alive instead of laying eggs as all other salamanders do.

The evolutionary process has produced myriad body shapes and sizes, but almost all salamanders have a similar appearance: elongate body, long tail, and relatively small legs compared with other terrestrial vertebrates. Salamanders worldwide have a body shape ranging from slender to robust, a distinct head, and four limbs, and the full range of aquatic and terrestrial body forms are found in the Southeast. Only the amphiumas, the sirens, and the European olm have an elongated, eel-like body with much reduced or lost limbs.

The smallest salamanders in the Southeast are the pygmy salamander (*Desmognathus wrighti*) and the patch-nosed salamander (*Urspelerpes brucei*), neither of which reaches a maximum total length of 2 inches as an adult. The largest species is the endangered Chinese giant salamander (*Andrias davidianus*) in east-central China, which may reach 6 feet in total length. The Japanese giant salamander (*Andrias japonicas*) is a close second. The related hellbender closely resembles these giants in body form and habits.

European fire salamanders (*Salamandra salamandra*) can squirt their poison at predators.

The Japanese giant salamander (*Andrias japonicas*) is closely related to the North American hellbender. It lives in streams and reaches a total length of nearly 5 feet. This individual was 4 feet 9 inches long.

SALAMANDER TAXONOMY

Taxonomy is the scientific field that deals with the classification and naming of plants, animals, and other organisms. Scientists engaged in this field classify and name salamanders in ways that reflect ancestral genetic relationships among species. Closely related species are placed within the same genus (plural = genera), and closely related genera are placed within the same family. All described organisms have a two-part scientific name, which is italicized to reflect its Latin or Greek origin. The genus name (e.g., *Plethodon*) is always capitalized. The second, or species, name (e.g., *cinereus*) is not. The example given here is the scientific name of the red-backed salamander, *Plethodon cinereus*. The name of the northern slimy salamander is *Plethodon glutinosus*, and that of the Shenandoah salamander is *Plethodon shenandoah*. The inclusion of these three species in the same genus indicates that they are more closely related than other species that are in the same family but in a different genus, such as *Eurycea* or *Pseudotriton*. Genus names are sometimes abbreviated after first usage if there is no chance of confusion (e.g., *P. cinereus*).

Taxonomy is not a static science. Early descriptions of new salamander species were based on the scientist's knowledge of the diversity of these animals in the region and recognition that the specimen in question was different enough to warrant a new name. Many southeastern salamanders were described in the 1800s and early 1900s in just this way. Molecular biology has revolutionized our understanding of the genealogies of species, genera, and families. Several techniques allow taxonomists to examine the genetic structure of species, leading to a much more refined understanding of the relationships of populations in different areas. These new techniques combined with improved accessibility to remote regions have resulted in descriptions of many new salamander species. The description process continues today and will continue into the near future as new species are discovered. The species accounts included here indicate which of the currently recognized species may actually include one or more new species and when each species was first described.

Modern genetic techniques have also produced questions about how "species" should be defined. Historically, a species has been defined as a group of interacting individuals and populations that can reproduce and leave viable offspring. Two or more distinct species living in the same place that do not hybridize are not able to produce offspring and are thus said to be "reproductively isolated." Groups of populations that we once thought were not isolated, however, may in fact be different enough to warrant rec-

The eastern red-backed salamander (top), the northern slimy salamander (middle), and the Shenandoah salamander (bottom) are all members of the genus *Plethodon*.

ognition at the species level. How different do populations have to be to deserve specific recognition? Scientists are still debating that question.

Some salamander species described on the basis of genetic differences are so similar in appearance to another species that they cannot be differentiated externally; only molecular techniques or the geographic location can identify them. Examples include the three species of gray-cheeked salamanders in the Blue Ridge Mountains of North Carolina, the Valley and Ridge salamander and the Shenandoah Mountain salamander in western Virginia, and the mountain dusky salamander group in the southern Appalachians. Fortunately, most salamanders have enough observable differences to distinguish them in the field.

Common, or standard, English names are not based on a set of rigid protocols as scientific names are. The common name of a species, particularly a widespread one, is likely to vary regionally. The spelling of a name, whether two words are hyphenated, and other details are hotly debated in some circles, although the differences are often very minor. Many people would like to see common names standardized as scientific names are, and so the debates continue. This book generally uses the scientific and standard English names in the most recent checklist of amphibians and reptiles published by the Society for the Study of Amphibians and Reptiles, which is written by groups of experienced scientists who represent all three national professional herpetological societies.

Did you know?

Salamanders usually have 19 pairs of chromosomes—compared with 24 in humans—but may have as many as 64 pairs. They have much more DNA than frogs.

Northern (top) and southern (bottom) gray-cheeked salamanders are closely related and are so similar in outward appearance that the only reliable way to differentiate between them is at the molecular level.

HOW TO IDENTIFY SALAMANDERS

Visually identifying and differentiating between most species of salamanders in the Southeast is relatively straightforward on the basis of body size, shape, color, and pattern. The fact that numerous species of terrestrial and stream-breeding salamanders have very well defined geographic ranges that do not overlap simplifies identification in some situations because all or most of the species that look similar can be eliminated from consideration on the basis of locality. Populations of some species show remarkable variation in color and pattern, and learning the range of variation in the salamanders in your area may be necessary to ensure accurate identification. For the majority of salamanders, however, a few particular characteristics can be used to determine the family and genus, and in many instances the species. Several basic features of appearance are commonly used in identifying most species of salamanders in the Southeast and differentiating them from other species that might be in the vicinity.

Body Size

Body size can be measured as the length of the body (snout to vent [the posterior margin of the cloacal opening]), as the body plus tail length, or as the weight of the body (mass). The maximum size of adults of a species is sometimes a useful character in differentiating among species that are similar in appearance, such as the dwarf and two-lined salamanders. Minimum and maximum lengths in this book are given as total length (body plus tail), so when you are using size to identify a species, it is important to consider whether an individual's tail has been broken. The adults of some salamanders are sexually dimorphic; usually the female is the larger sex.

Salamander biologists sometimes draw blood for use in genetic analyses to determine relationships among salamanders and to identify some species.

Body Shape

All salamanders have a basic elongated shape, but species can be characterized as slender or stout, and some have a flattened head and body. The proportional tail length can also be a distinctive feature; the long-tailed salamander, for example, has a tail that is nearly two-thirds of the total body length. The shape of the tail varies among species from the oarlike tail of fully aquatic sirens and mudpuppies to the tails of some dusky salamanders that are round in cross section at the base. Once you are familiar with all the salamanders in your area, you will find that identification can often be made on the basis of shape alone.

Legs and Toes

The absence of hind legs immediately differentiates the sirens and dwarf sirens from all other salamanders. The remaining species of salamanders all have four legs, but in proportion to their body size. The tiny legs of amphiumas are distinctive enough to keep any species in this genus from being confused with any other group of salamanders. The number of toes on front or hind feet can also be distinctive; the amphiumas, for example, have one, two, or three toes per foot depending on the species. Four toes on each back foot rather than the five present in most salamanders are characteristic of the dwarf salamanders, four-toed salamander, waterdogs, and mudpuppies.

Costal Grooves

Vertical grooves, which are visible along the sides of most salamanders, indicate the number of vertebrae and are an effective way to identify some species. For example, a mole salamander may have 10 costal grooves, many

The absence of rear legs immediately distinguishes the sirens and dwarf sirens (such as the individual shown here) from all other salamanders, including the similar amphiumas.

The vertical indentations between the legs of many salamanders correspond with the number of vertebrae in the backbone and are used in some cases to distinguish between species.

of the dusky salamanders have 14, and greater sirens have as many as 40. However, newts, waterdogs, and mudpuppies have no costal grooves.

Genetics

Many species of salamanders, especially within the family Plethodontidae, are "electromorphs"; that is, they differ genetically but are identical in appearance. Even salamander biologists cannot distinguish between some species without using genetic analysis techniques. The exact location of capture can help in determining the identity of most species that look similar, but not all. Three factors make identification difficult in such cases. First, many closely related species interbreed in their zones of overlap, resulting in hybrids that have characteristics of both species. The occurrence of hybridization among various forms is noted in the species accounts. Second, two or more salamander species may be virtually identical in body shape, color pattern, habitat, and behavior, but because they differ in their genetic makeup they qualify as separate species. Thus, an individual salamander belonging to one of the 10 species of slimy salamanders that occur in the Southeast may not differ appreciably in appearance from Virginia to Florida to Louisiana. Third, individuals of some species exhibit a range of body colors and patterns within the same population. The Ocoee salamander, for example, may be nearly black with a chestnut spot pattern on the back or may be orange without distinct markings—or may exhibit some variation of these characters. The best approach is to learn how the populations and species in your area vary in color and pattern.

Body Markings

Particular colors and patterns on the head, body, legs, and tail are characteristic of many species of salamanders. The presence, position, width, shape, and intensity of dark or light stripes down the back and onto the tail are key features in the identification of two-lined, three-lined, dwarf, and zigzag salamanders; striped newts; and others. The size and pattern of dark or light spots can often be used to distinguish between mud salamanders and red salamanders or between slimy salamanders and some of the other woodland species. Individuals within the same population of eastern or southern red-backed salamanders may have a prominent reddish stripe down the back or no stripe at all. The two sexes of the marbled salamander differ subtly in their black and white coloration, with the males being a brighter white.

Special Head Characters

Several head, jaw, and chin features are characteristic of some species in the family Plethodontidae, and may even classify an individual salamander as belonging to a particular genus.

NASOLABIAL GROOVE The presence of a thin groove that runs from the nostril to the upper lip and is visible to the naked eye is a signature trait of members of the family Plethodontidae. Although not useful for distinguishing among species within the family, the presence or absence of the nasolabial groove can readily be used to separate plethodontid salamanders from mole salamanders and newts in the Southeast.

CIRRI Some species, such as the southern two-lined salamander (genus *Eurycea*), have a pair of fleshy projections that point downward from the upper lip. Cirri are most prevalent in males during the breeding season but are present in females of some species.

Downward projections (cirri) in male plethodontid salamanders appear in the breeding season and are likely used for stimulation of and communication with females.

MENTAL GLAND An enlarged round or oval disk called the mental gland is present beneath the chin of males in the genus *Plethodon* during the breeding season. The mental gland is more prevalent in some *Plethodon* species than others. It is absent in salamanders in other genera, with the exception of the pygmy salamander, which also has a mental gland.

During the mating period the mental (or hedonic) gland on the chin of male plethodontids, such as this Blue Ridge gray-cheeked salamander, enlarges. The male is on the left and the female is on the right.

Spring salamanders have a distinct ridge bordered by a light line along each side of the snout between the eye and nostril.

All members of the genus *Desmognathus* have a light line—the eye–jaw line—that runs from the eye to the back of the jaw.

CANTHUS ROSTRALIS The spring salamander has a distinctive bony ridge known as the canthus rostralis that runs from the eye to the nostril. The ridge is usually marked by a dark-bordered white line.

EYE–JAW LINE Dusky salamanders can usually be distinguished from other species in their range by an obvious light line that runs posteriorly from the eye to the jaw.

TEETH Tooth shape is important in distinguishing some closely related species. For example, the northern dusky salamander has narrow crowns on its teeth, whereas its close relative the Virginia dusky salamander has relatively wider crowns.

Skin Texture

Most salamanders have soft skin that can range from the rubbery feel of the mole salamanders to the mucus-covered bodies of sirens, mudpuppies, and amphiumas. Newts in the terrestrial eft stage have granular skin that differentiates them from other forest-dwelling salamanders in the Southeast.

External Gills

More than 60 species of southeastern salamanders are permanently aquatic or have aquatic larvae. Most of them have external gills that are visible outside the body, although adult amphiumas and hellbenders have gill openings but no visible gills. The more than 40 woodland species that are terrestrial breeders bypass the typical larval stage and develop completely within the egg. The hatchlings have no external gills, although gill buds—remnants of the embryonic gills—are sometimes present.

Environmental Information

The environmental conditions under which a salamander was found can often help in identifying it. For example, knowing that marbled salamanders emerge from underground retreats and breed in the first rains of the fall will help to distinguish them from tiger or spotted salamanders, which breed months later.

Geographic Location

The geographic location where a salamander is found in the Southeast can sometimes provide a clue to its identity by eliminating similar species not found there. A waterdog from a river in southern Alabama, for example, can be identified as a Gulf Coast waterdog simply on the basis of where it was found.

Habitat

The specific habitat, such as aquatic or terrestrial, stream or pond, forest or clearcut, can be very helpful in species identification. The species accounts indicate habitats where species are most likely to be found.

Time of Day and Year

Some southeastern salamanders migrate during their breeding season and are more likely to be found in some areas than in others. In the case of some woodland salamanders, the movement is vertical, from the surface to deeper underground. Webster's salamanders are highly unlikely to be discovered beneath a log or rock in July or August when they have retreated underground, although they may be common at such sites in January or February. Different species of seasonal wetland–breeding salamanders can be expected to be moving at night during rains at specific times of the year. With the exception of newts, which are occasionally seen moving about during the day, salamanders characteristically move about above ground on humid or rainy nights.

The external gills of Tennessee cave salamanders are used for respiration. The rows of light dots on the head and body of this individual are part of the lateral line system used to detect movement through water.

Georgia blind salamanders are completely aquatic and are found in limestone caves, sinkholes, and artisan springs.

Salamander Habitats in the Southeast

Salamanders of the Southeast live in a variety of aquatic and terrestrial habitats. Some species are strictly aquatic and never go on land; others are strictly terrestrial. Many species require both types of habitat in order to complete their life cycle. Species' adaptations to one or more of these habitat types have led to natural partitioning into ecological groupings that generally, although not always, follow family lines.

All southeastern species in the families Amphiumidae (3 species), Cryptobranchidae (1 species), Proteidae (5 species), and Sirenidae (4 species) are permanently aquatic and seldom if ever venture onto land. Five of the 77 southeastern species in the family Plethodontidae are aquatic as well, while 41 species are terrestrial in all stages of their life cycle and are found exclusively in mixed hardwood or spruce-fir mountain forests. All 10 of the Ambystomatidae species found in the Southeast, 2 species of newts (family Salamandridae), and the other 34 species of plethodontids live in or around forest habitats but breed in aquatic areas. Some lay their eggs in small streams or seepage areas, others in isolated wetlands.

Some habitats have a greater abundance and diversity of species than others. Because salamanders are so seldom seen, however, the high number of species and individuals may go unnoticed by anyone but a herpetologist. Knowing the general habitat types favored by salamander species as well as the microhabitats where they are most likely to be found is the first step toward enjoying a group of fascinating creatures that few people ever see.

Did you know?

Climate change can cause loss of isolated wetlands and declines of pond-breeding salamanders in parts of the Southeast. Reduced rainfall and longer intervals between rains decrease the length of time ponds hold water and prevent larvae from reaching metamorphosis.

Southern dusky salamanders live in rocky streams in mountainous regions.

FRESHWATER WETLAND HABITATS

The aquatic habitats of the Southeast can be categorized on the basis of several characteristics. Seasonal variation in size and permanence, annual extremes and variation in temperature and rainfall, water chemistry, and the presence or absence of other species—especially fish—are all factors determining which salamanders will be found in a body of water. The following generalized categories pertain to identifiable aquatic habitats with which various salamander species are associated.

Ponds, Lakes, and Reservoirs

Permanent lakes and ponds, such as those formed by beavers damming small streams or those created where the earth collapses in some limestone areas to form sinkholes, are natural habitats in the Southeast. Man-made farm ponds, which may function much like beaver ponds, are also liberally distributed throughout the region. Large, deep, natural lakes of glacial origin are absent from the Southeast, but artificial reservoirs created by damming large rivers are physically similar. A few salamander species are associated with these habitats, but except for the mudpuppy, which can survive in large lakes where fish are present, the eggs, larvae, or even the adults of most end up as prey. Salamanders lay their eggs along the flooded peripheries of permanent ponds, lakes, and reservoirs where emergent vegetation is present and the water is shallow enough to preclude the presence of predatory fish.

Natural ponds such as this Carolina bay support populations of mole salamanders, sirens, amphiumas, and newts.

Isolated Seasonal Wetlands

Isolated seasonal wetlands are found throughout the Southeast. They come in different shapes and sizes and have several geological origins. The Carolina bays common in much of the Coastal Plain from southeastern Georgia to Virginia, vernal ponds, sinkholes, lime sinks, and scoured depressions in stream or river floodplains have basic features that make them ideal for certain groups of salamanders. Many dry up in some years, and some dry up every year, so they are unlikely to have fish that prey on salamanders or their eggs. Some species of mole salamanders such as the marbled and spotted salamanders require isolated fish-free wetlands for breeding success. Salamanders can breed in seasonal wetlands that vary in surface area from smaller than a backyard to several acres. The depth may also vary, from a few inches in some to more than 10 feet in others. Among the factors determining how long an isolated wetland holds water during a year are the porosity of the substrate (e.g., some Carolina bays have a clay lens that helps retain standing water), the amount and timing of rainfall, evaporation rates (which in turn depend on air temperature and humidity), transpiration rates (which depend on the amount and type of standing vegetation and how much water is released through the leaves to the atmosphere), and proximity to streams or rivers that might flood.

Vernal ponds (top left) are seasonal wetlands that often dry during late summer and are fish-free during the period when many of the mole salamanders breed.

Tiger salamanders and other mole salamanders have been observed breeding in ponds (bottom left) with ice on the surface.

Sinkhole ponds and isolated ponds (right) created by humans are used extensively by pond-breeding salamanders and newts.

Bogs with shallow water and soft, mucky substrates (left) are usually spring fed and support four-toed salamanders, mud salamanders, and red-spotted newts.

Mudpuppies and hellbenders are found in cold, shallow rocky streams (right).

Regions of the Southeast where isolated wetlands are scattered abundantly across the landscape are home to numerous forest-dwelling salamander species that breed aquatically, including mole salamanders, newts, and dwarf salamanders.

Bogs

Mountain bogs in the Blue Ridge Mountains, pitcher plant bogs in the Coastal Plain, and wetlands where small streams meander through old pastures are usually small, acidic, and fish-free. Newts, four-toed salamanders, and mud salamanders sometimes breed in these wetland habitats. One-toed amphiumas can be found in pitcher plant bogs.

Rivers, Large Streams, and Spring Runs

The rivers and large streams of the Southeast and the clear limestone springs and streams of Georgia and Florida are impressive aquatic habitats that set the cultural character and often shape the natural biological features of a region. But aside from marginal areas of flooding such as river swamp bottomlands, they are not home to many salamanders. Hellbenders inhabit mountain streams with large rocks and riffles. Sirens and amphiumas, which reach sizes large enough to deter most fish predators, are sometimes found in Coastal Plain rivers and streams. The five species of permanently aquatic waterdogs inhabit streams and rivers from Virginia to Louisiana in different parts of the Coastal Plain and Piedmont.

Floodplain River Swamps, Oxbow Lakes, and Bottomland Swamps

The seasonal drying and flooding regimes of floodplain river swamps, backwater oxbow lakes, and bottomland swamps are particularly dependent on local and upstream rainfall. The extensive floodplain swamps that border the slow-moving rivers and large streams of the southeastern Coastal Plain are breeding sites for newts and several species of forest-dwelling mole salamanders. Plethodontid species such as slimy salamanders, three-lined salamanders, and red salamanders may be present in the surrounding swamp forests.

Small Streams and Seepage Areas

Small, rocky streams formed from mountain springs and watershed runoff are habitats with high salamander diversity and abundance, as are the wetland habitats known as seeps. Seeps can be created by groundwater slowly flowing out at the base of a steep bluff or between bedrock formations. Little rivulets less than a foot wide and only a few inches deep may flow away from a bluff, sometimes accumulating into small streams that flow into larger ones. Leaf litter and debris are common characteristics of such sites. Dusky, two-lined, dwarf, mud, and red salamanders are often abundant in seeps and may be found hiding in and under the debris. Rocky streams flowing through forest habitat, even small ones whose flow is noticeable mostly during wet periods, may still have high numbers of stream salamanders such as black-bellied, spring, dusky, and Junaluska salamanders.

Small, rocky streams (left) throughout the Southeast are habitat for a variety of small salamanders with an aquatic larval stage.

Two-lined and three-lined salamanders, red and mud salamanders, and many of the dusky salamanders thrive in stream or seepage areas (right) within hardwood forests.

TERRESTRIAL HABITATS

The terrestrial habitats of the Southeast can be categorized according to their vegetation cover, soil type, topography, and geological history. The following generalized categories describe identifiable terrestrial habitats with which various salamander species are associated.

Mixed Hardwood Forests

Hardwood forests, which occur throughout the Southeast, can include or may even be dominated by a broad array of such tree types as oaks, hickories, magnolias, maples, and gums. Many mixed hardwood forests have interspersed pine trees. In their natural state, these forests offer vital hiding places for salamanders beneath the ground litter of decaying leaves, dead bark, and logs; and their thick canopy provides shade and helps to retain moisture. More than two-thirds of the salamander species in the Southeast can be found in mixed hardwood forests. Adult mole salamanders and the eft stage of newts spend most of their lives in such forests, but always within migrating distance of wetland breeding habitats. Two-lined salamanders, red salamanders, and many of the dusky salamanders depend on healthy hardwood forests being adjacent to the stream or seepage areas where they breed and spend much of their time. Most woodland salamanders (genus *Plethodon*) are terrestrial breeders that live their entire lives in mixed hardwood forests, sometimes on the surface and sometimes in burrows, crevices, or other deep underground retreats.

Pine Forests

Several species of pine trees can be found scattered among hardwood forests at all elevations and under differing soil and moisture conditions, but pine forests abound in the southeastern Coastal Plain. Most of our longleaf pine–wiregrass forests are mere remnants of the once vast fire-maintained savannas that covered the Coastal Plain from Virginia to Texas. Even the dry longleaf pine forests of the sandhills can support a variety of forest-dwelling salamanders such as tiger salamanders, which survive in underground burrows and travel to seasonal wetlands for breeding. The low-lying pine flatwoods habitat found closer to the coast once dominated large areas of the Southeast. Pine flatwoods, which consist primarily of longleaf and slash pines, once made up almost half of Florida's natural habitats. Seasonal wetlands in these habitats serve as breeding sites during the winter for striped newts and federally protected flatwoods salamanders.

Although they are amphibians, some southeastern salamanders are strictly terrestrial and do not require ponds or other wetlands for survival. Logs and other surface objects in terrestrial habitats are used extensively by several salamanders.

Many forests in the southeast are composed of a mix of hardwood and pine. Fewer species occur here due to the drier substrate.

Although longleaf pine forests and flatwoods occur on sandy and xeric soils, several southeastern salamanders are able to find shelter in these upland habitats.

A stream flowing through a cave can be an ideal habitat for several species of salamanders, some of which may be permanently aquatic.

Agricultural landscapes are usually inhospitable to salamanders, but these amphibians will cross long distances to reach wetlands for breeding.

Spruce-Fir Forests

The steep, rocky, mountainous terrain at higher elevations in the Appalachian Mountains of Virginia, North Carolina, and Tennessee, including Great Smoky Mountains National Park, is the home of coniferous forests dominated by spruce and Frasier fir. The heavy layer of moss coupled with the extensive rocky latticework of underground hiding places and continually moist conditions form ideal habitats for terrestrial-breeding species such as red-legged and pygmy salamanders.

Limestone Caves

Common features of the landscape in areas of karst topography are caves and crevices created as the limestone dissolved away over the eons. In addition to the cave salamander, several forest-dwelling species—including Pigeon Mountain, Cumberland Plateau, long-tailed, and zig-zag salamanders—will enter caves, sometimes for weeks or months at a time during periods of drought or extreme cold or heat. The Georgia blind salamander, Tennessee cave salamander, and Berry Cave salamander are found only in caves and live permanently in pools associated with underground streams and rivers.

Some woodland salamanders are associated with limestone rock outcrops and disappear deep underground during hot, dry weather.

Urban, Suburban, and Agricultural Habitats

Salamanders differ from other major groups of vertebrates—including frogs, snakes, lizards, and turtles—in being unable to thrive as well in areas developed by humans as they do in natural areas. A few species can persist for long periods of time in urban parks, suburban areas where small wooded tracts are left intact, and even on golf courses if fish-free wetlands are present. Tiger salamanders sometimes burrow in agricultural fields, but they must also travel to aquatic breeding sites and therefore may have to survive long-distance migrations that involve crossing busy highways. Two-lined salamanders can tolerate degraded streams in urban zones.

Frosted flatwood salamander populations have declined dramatically in the past two decades. This species is on the Endangered Species List maintained by the U.S. Fish and Wildlife Service.

Species Accounts

ORGANIZATION AND ORDER OF SPECIES ACCOUNTS

The names used in the species accounts—both the scientific name and the common name—are those generally accepted by salamander biologists in 2008. The Taxonomy section in each account lists alternate names that have been used or proposed for the species. Each species account includes a description of adults, larvae, and juveniles, as well as other salamander species that are similar in appearance. The accounts describe habitat, behavior, activity patterns, reproduction, diet, and predators. Each account also includes a discussion of relevant conservation issues.

In contrast to other amphibians (frogs) and reptiles (snakes, turtles, lizards) of the Southeast, a high proportion of our salamanders have highly localized distribution patterns, sometimes being known from only a single mountain in one or a few counties. The regional distribution in the Southeast, as well as the U.S. distribution of certain species, is presented on maps that accompany each account.

The accounts are grouped into three major assemblages: aquatic, semi-aquatic, and terrestrial. They are further partitioned into categories based on salamander life-cycle patterns and the habitats with which species are most likely to be associated. Some salamander species retain their larval gills and remain aquatic their entire lives. Many salamanders, such as the

Spring salamanders are the largest salamander species in mountain streams, where they prey on other salamanders.

stream salamanders, are semiaquatic, breeding in water but venturing into the surrounding forest during wet spells. Woodland salamanders (genus *Plethodon*) do not require water at all for breeding and larval development. They lay their eggs in moist places under rocks or in logs, and development takes place inside the egg capsule. The hatchling of such species is a miniature of the adult body form. These terrestrial-breeding salamanders exploit the moist, shaded environment of hardwood forests. *To see the order of the species accounts at a glance, consult the table on pages 306–7.*

AQUATIC SALAMANDERS Fifteen species of southeastern salamanders live in rivers, streams, swamps, lakes, or ponds. The diversity of these fully aquatic salamanders is greater in the Southeast than anywhere else in the world.

BASIC FEATURES OF THE SPECIES ACCOUNTS

Webster's salamanders inhabit rocky areas in hardwood forests with heavy ground litter. The species is most likely to be seen on the surface from late spring to early summer.

Quick identification guide

How do you identify a Webster's salamander?

BODY SHAPE
Small and slender

NUMBER OF COSTAL GROOVES
17–19

BODY COLOR
Brown or reddish brown

PATTERN
Usually a broad stripe down the back and onto the tail

BODY SIZE
Adults to 3.3 inches in total length

Webster's Salamander   *Plethodon websteri*

DESCRIPTION Webster's salamanders are tiny brown salamanders that range in color from reddish to yellowish brown. Some individuals have a stripe down the back to the end of the tail, and a small proportion of the population has no stripe at all. The sides of the stripe are wavy and are usually brighter on the front part of the tail than on the body. The sides have lighter-colored flecks, and the belly is mottled with orange, red, black, and white.

LARVAE AND JUVENILES Recent hatchlings are about 0.5 inch long and resemble adults in coloration.

SIMILAR SPECIES Webster's salamanders can be confused with zigzag salamanders, but the two species' ranges overlap in only a few areas. Because both species are variable in color pattern, the two must be distinguished by genetic analysis.

DISTRIBUTION Webster's salamanders are found from South Carolina to Louisiana, but except for a relatively broad area from western Georgia through east-central Alabama they occur as isolated populations.

HABITAT Hardwood forests in rocky terrain with heavy ground litter of leaves, logs, and rocks are the preferred habitat.

278 • *Webster's Salamander*

AQUATIC CAVE SALAMANDERS Three species of southeastern salamanders are completely adapted to life in the total darkness of caves.

STREAM AND SEEP SALAMANDERS Of the species of salamanders that are at least partially terrestrial, 31 in the Southeast breed and stay in the vicinity of small streams or seepage habitats in wooded areas.

SEASONAL WETLAND SALAMANDERS Thirteen species of terrestrial salamanders live in forests but move overland during their annual breeding season to isolated seasonal wetlands such as Carolina bays and vernal ponds.

FOREST TERRESTRIAL SALAMANDERS These fully terrestrial species do not require water in which to breed and have no larval stage. The 41 species that occur in the Southeast constitute a remarkably high species diversity.

BEHAVIOR AND ACTIVITY These salamanders are locally abundant on the surface beneath ground litter from October to May but virtually disappear deep underground in summer through early fall.

FOOD AND FEEDING Webster's salamanders eat small terrestrial invertebrates such as ants, mites, termites, springtails, spiders, snails, centipedes, flies, and worms.

REPRODUCTION Terrestrial courtship and mating occur from January to March. Females begin laying their clutch of 3–8 eggs underground in June and July. Hatchlings are found above ground in October.

PREDATORS AND DEFENSE Specific predators have not been documented but presumably include ringneck snakes and garter snakes, large spiders, and shrews that occupy similar habitat.

CONSERVATION No generalized conservation threats have been identified for Webster's salamander other than loss of forested habitat. South Carolina lists the species as Endangered, and Louisiana considers it a Species of Special Concern.

TAXONOMY Populations of Webster's salamanders were reported in the scientific literature as early as the 1950s, but as zigzag salamanders. The species was described in 1979, and no subspecies have been recognized.

Did you know?

New species of salamanders continue to be described in the Southeast as DNA analyses differentiate species that look identical. Several new species are expected to be described within the next several years, even as the actual number of salamander populations continues to decline.

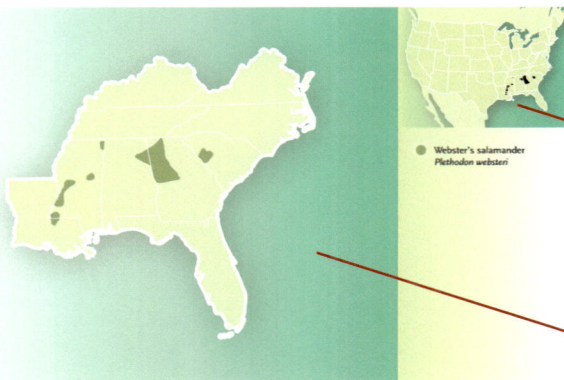

Webster's salamander
Plethodon websteri

Natural history information along with interesting or anomalous facts about the species or about salamanders in general

Small map with a shaded area showing where the species occurs in the continental United States

Large map showing where the species occurs in the Southeast

AQUATIC SALAMANDERS

Two-toed amphiumas are the longest salamanders in North America. The record total length is more than 45 inches.

How do you identify an amphiuma?

Amphiumas

One-toed Amphiuma	*Amphiuma pholeter*
Two-toed Amphiuma	*Amphiuma means*
Three-toed Amphiuma	*Amphiuma tridactylum*

DESCRIPTION Adults are larger than most other salamanders, and are robust and cylindrical with tiny eyes and four very small limbs. External gills are lacking, but a single pair of gill slits is present. One-toed amphiumas are fairly uniform in body coloration and have a single toe on each of the four tiny legs. Two- and three-toed amphiumas are uniformly dark gray or brown to nearly black with a light gray belly. The body and belly colors are more distinct in the three-toed amphiuma, which also has a dark throat patch that is not present in the other two species. Two-toed amphiumas have two toes on each leg, and three-toed amphiumas have three toes.

Adult amphuima have a single pair of gill slits, tiny eyes, and four very small limbs.

BODY SHAPE
Long and eel-like; four small legs

NUMBER OF COSTAL GROOVES
None

BODY COLOR
Dark brown or gray to nearly black, with or without gray belly

PATTERN
Uniform

BODY SIZE
Adults to 13 (one-toed), 45 (two-toed), and 41 (three-toed) inches in total length

The contrast between upper and lower body colors is more distinctive in the three-toed amphiuma.

Hatchling and juvenile amphibians (left) have external gills that are absorbed soon after hatching.

The one-toed amphiuma's characteristic gill opening and single toe are apparent on this specimen (right). The lateral line system, visible in slits on the snout and in pores on the body, senses vibrations in the water.

LARVAE AND JUVENILES Larval amphiumas resemble adults in body form and are black with a light gray belly; they have white gills that are lost shortly after hatching. Juveniles have the same color and pattern as adults.

SIMILAR SPECIES Greater and lesser sirens are long and cylindrical but have large external gills and only one pair of small limbs, which are near the head.

DISTRIBUTION One-toed amphiumas are found in southern Mississippi, Alabama, Georgia, and from the Florida panhandle south in the Gulf coastal lowlands to Hernando County. Two-toed amphiumas occur in the Coastal Plain from Virginia south through peninsular Florida and west to eastern Louisiana. Three-toed amphiumas occur from southeastern Arkansas to the Gulf of Mexico and from central Alabama to eastern Texas.

HABITAT Adult and juvenile one-toed amphiumas occupy swampy and periodically inundated floodplains associated with streams having low to moderate flow. They are also found in mixed bottomland hardwoods and cypress habitats with deep muck derived from hardwood and cypress litter, and in pitcher plant bogs. Two- and three-toed amphiumas are generally

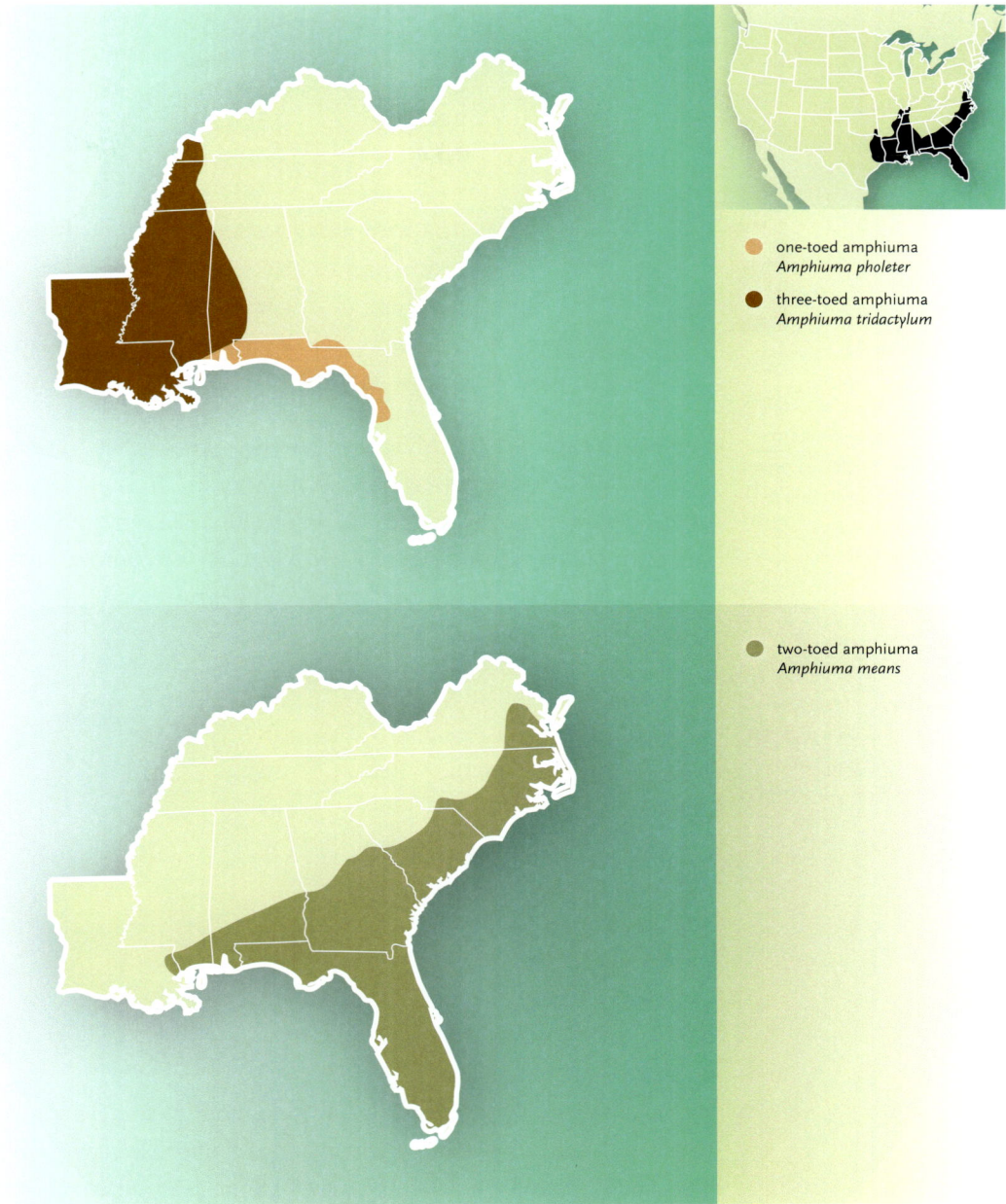

one-toed amphiuma
Amphiuma pholeter

three-toed amphiuma
Amphiuma tridactylum

two-toed amphiuma
Amphiuma means

Like the other two species, two-toed amphiumas are fully aquatic predators of a wide array of invertebrates and small vertebrates. They can inflict a nasty wound if they bite you.

found in hardwood swamps, cypress swamps, temporary ponds, permanent ponds and lakes, sloughs, canals, and sluggish streams; they sometimes occur in clear, sand-bottomed creeks with little or no detritus on the bottom. They may seek shelter in burrows in the bank.

BEHAVIOR AND ACTIVITY All life history stages are fully aquatic. Amphiumas seldom leave the water but will move short distances over land during heavy rains. They can survive droughts and dry ponds by aestivating in burrows in the substrate. Activity decreases during cold weather.

FOOD AND FEEDING Hatchlings and juveniles probably feed on small invertebrates, but little is known about how they capture and handle prey. Adults of the two larger species are powerful predators. Collectively, the three species of amphiumas have been reported to eat crayfish, worms, many insects and their larvae, small frogs and tadpoles, other salamanders, watersnakes, turtles, lizards, small fish, snails, and spiders.

Females lay fewer than 200 eggs annually and, like this three-toed amphiuma, live for many years.

REPRODUCTION Courtship and mating are completely unknown in the one-toed amphiuma and have not been thoroughly described for the two- and three-toed amphiumas. Females lay strings of 20–200 eggs beneath objects such as logs and other debris in winter and

spring, and usually remain coiled around them during incubation. Eggs have been found beyond the water's edge, but these may have initially been laid in water before the water level dropped. Eggs hatch in about 5–6 months; the larval stage lasts about 3 weeks.

PREDATORS AND DEFENSE Amphiumas are the primary prey of adult mud snakes, but cottonmouths, kingsnakes, some watersnakes, alligators, wading birds, sandhill cranes, and otters will eat them as well. Feral hogs rooting in the mud of drying ponds and lakes dig amphiumas out of the bottom and may constitute a major source of mortality. The larvae and small juveniles are potential prey of predatory insects, small aquatic snakes, and

Individuals of all three species of amphiumas will bite one another during mating rituals and in male-male aggression. This one-toed amphiuma shows bite-mark scars from such encounters.

fish. Large amphiumas bite hard and can inflict a serious wound on potential predators, including humans. They also produce slimy mucus that makes them very slippery.

CONSERVATION The one-toed amphiuma is rare throughout its range and is of conservation concern in all states where it occurs; Alabama considers it a Species of High Conservation Concern. Two- and three-toed amphiumas are widely distributed, often in large populations, and are not of serious conservation concern in most areas. Feral hogs, an introduced species, may consume significant numbers of amphiumas when ponds and lakes are drained, and the effects of their predation deserve further examination.

TAXONOMY Three species of amphiumas are recognized in the Southeast. The one-toed amphiuma was not described until 1964. Two-toed amphiumas were described in 1821; three-toed amphiumas were described in 1827. No subspecies are recognized. Congo eel and lamp eel are common names in some regions.

Did you know?

Amphiumas can migrate from one body of water to another during heavy rains by moving across wet land and in very shallow water.

This hellbender in a mountain stream has the species' characteristic broad, flattened head and body, paddle-shaped tail, and wrinkled folds of skin on the sides.

How do you identify a hellbender?

BODY SHAPE
Large and flattened top to bottom

HEAD SHAPE
Broad and flattened with small eyes

BODY COLOR
Light brown, orangish, or nearly black

PATTERN
Faded dark spots or blotches on a lighter background

BODY SIZE
Adults to 29 inches maximum, but most are a little over 12 inches long

Hellbender

Cryptobranchus alleganiensis

DESCRIPTION The fully aquatic adult is large and flattened with wrinkled folds of skin on each side of the body. The head is very broad and also flattened, and the eyes are small. The paddle-like tail is broad and flat. Hellbenders can range from brownish to bright orange; some have small spots or irregularly shaped blotches. The belly is usually plain.

LARVAE AND JUVENILES Larvae are chunky, have bushy gills, and develop irregular spots as they mature. They absorb their external gills after about 2 years when they reach 4–5 inches in length, and attain maturity within 5–8 years. Juveniles resemble adults in color and pattern.

SIMILAR SPECIES The broad, flat body of hellbenders readily differentiates them from mudpuppies, the only other large salamanders whose geographic range overlaps theirs. Mudpuppies also have obvious fluffy gills as adults.

DISTRIBUTION Eastern hellbenders occur in northern Alabama, Mississippi, and Georgia, much of Kentucky and Tennessee, western North Carolina, and southwestern Virginia. They range up the Appalachians through Pennsylvania into New York. With the exception of a few streams in the Northeast that are part of the Chesapeake Bay watershed, they occur only in drainages that eventually empty into the Mississippi River.

HABITAT Hellbenders occupy rivers and large streams, and also small mountain streams as long as these are permanent. They prefer riffle zones with large, flat rocks but may be in any portion of the stream that offers cover.

BEHAVIOR AND ACTIVITY These impressive salamanders rarely leave the water. They emerge from their rocky hiding places at night to search for food, but they can be very active in the daytime during the breeding season (September and October). In some streams they are diurnal during much of the summer. Males are territorial during the breeding season and defend nesting rocks from other hellbenders, but even outside the breeding season a single rock seldom shelters more than one. Hellbenders generally move only short distances and maintain small territories of a few hundred square feet, but they have been known to move several miles within a river. They are very slimy and therefore very difficult to catch and hold. The natural life span may be 30 years or more.

Larvae (left) have bushy gills and develop irregular spots as they mature.

Hellbenders' sharp, serrated teeth (right) can result in a nasty bite.

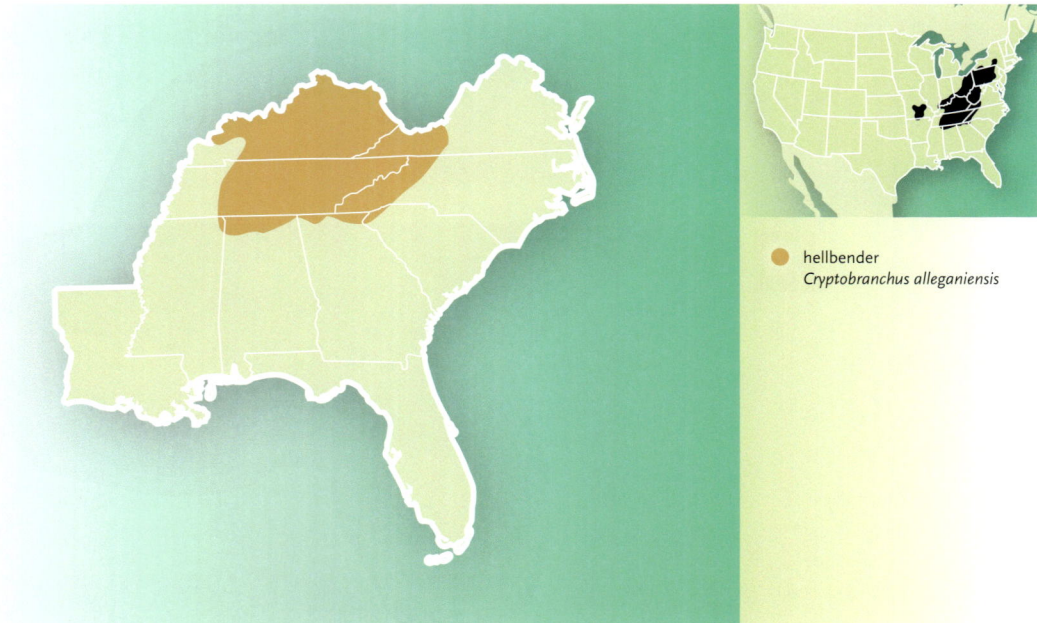

● hellbender
Cryptobranchus alleganiensis

FOOD AND FEEDING Hellbenders eat crayfish primarily, and also small fish, aquatic insects, salamanders, and occasionally other hellbenders. Game fish do not appear to be a typical food item.

REPRODUCTION Females lay 200–400 or more eggs, usually between August and October, in depressions the male makes under his rock. A male may mate with and fertilize the eggs of several females. The male guards the eggs until they hatch some 2 months later. Some males will allow their young to remain under their rock for 6 months or more. Nesting habitat (large rocks not covered in silt) seems to be a limiting resource in most streams.

Hellbenders may be caught by fishermen. They are often killed as a result or let go with the hooks still in their mouths.

PREDATORS AND DEFENSE The common enemies of adults include humans, who hook hellbenders on lines while fishing or trapping and kill them. Large fish, some turtles, and northern watersnakes eat hellbenders; otters probably are natural predators as well. Larvae and small juveniles are presumably eaten by large predatory insects and fish as well as other hellbenders. These rather intimidating salamanders sometimes bite if handled but are *not* venomous. Their main defenses against predators are concealment under rocks, their camouflage, and the noxious and distasteful mucus that covers their body and makes them difficult to catch and handle.

Without the normal color and pattern to provide camouflage, albino hellbenders are more easily seen and captured by predators.

Hellbenders generally move only short distances and maintain small territories of a few hundred square feet, but they have been known to move several miles within a river.

CONSERVATION Hellbenders no longer occupy much of their historic range because siltation from land erosion has destroyed their habitat. Damming of rivers has also eliminated much of the hellbender's original habitat, especially in the Tennessee River valley. Forest cover along streams and rivers is important for maintaining healthy water for hellbenders. The hellbender is a Species of Special Concern in North Carolina and Virginia, is listed as Threatened by the state of Georgia, and is deemed in need of management in Tennessee. It is very rare in Alabama, where it is classified as a Species of Highest Conservation Concern. Anglers can affect local hellbender population levels in any stream by deciding whether to kill these salamanders or release them. Habitat management programs to conserve and protect the rivers and streams in which this unique, long-lived salamander occurs are needed for conservation efforts. Accurate data on population sizes throughout the geographic range would allow a better evaluation of the species' conservation status.

TAXONOMY The eastern hellbender (*Cryptobranchus alleganiensis alleganiensis*) was first described in 1803. Other common names are grampus, alligator, snot otter, devil dog, mud devil, mollyhugger, and giant salamander. Hellbenders are also often erroneously called mudpuppies or waterdogs.

Did you know?

The Ozark hellbender, a western subspecies of the eastern hellbender, declined precipitously during the late twentieth century as a result of habitat alteration related to human recreation, siltation, pollution, and overcollection.

Shovel-nosed salamanders inhabit cool, rocky streams where their dull brown and yellowish body pattern serves as excellent camouflage.

How do you identify a shovel-nosed salamander?

BODY SHAPE
Head flattened top to bottom; tail flattened side to side

NUMBER OF COSTAL GROOVES
14

BODY COLOR
Dark brown, gray, or black

PATTERN
Two rows of yellowish brown, gray, olive, or yellowish spots

BODY SIZE
Adults to 5.75 inches in total length

Shovel-nosed Salamander

Desmognathus marmoratus

DESCRIPTION Shovel-nosed salamanders are large and robust. Their common name comes from the slanted, wedge-shaped snout. The slant starts well behind the small eyes on the flattened head, while the slope in other salamanders found in the same region starts in front of the eyes. Body coloration consists of two rows of highly irregular yellowish brown, gray, olive, or yellowish spots over a dull, dark brown, gray, or black background that provides excellent camouflage in the rocky streams in which this salamander lives. Two rows of tiny light spots occur on each side of the body in some populations. The belly is gray and may either lack a pattern or have dark mottling; it darkens in the center with age. The tail is flattened side to side and has a sharp upper keel.

A slanted, wedge-shaped snout gives the shovel-nosed salamander its name.

LARVAE AND JUVENILES Hatchlings have two rows of light spots on a gray back. They are nearly black and have whitish gills. Other *Desmognathus* larvae that co-occur with them are not as dark. Older juveniles are lighter

shovel-nosed salamander
Desmognathus marmoratus

Shovel-nosed
salamander larva

and have noticeable light flecks on each side, and their legs are longer than those of adults.

SIMILAR SPECIES The black-bellied salamander may be confused with the shovel-nosed, but the former's uniformly black belly, bulging eyes, and more conspicuous eye-to-jaw stripe are distinctive. All other members of the genus *Desmognathus* have two round openings inside the roof of the mouth (internal nares); shovel-nosed salamanders have only slits. Although they may be difficult to see, the internal nares may be visible in live salamanders if the mouth can be pried open carefully.

DISTRIBUTION Shovel-nosed salamanders occur in the southern Appalachian Mountains from southwestern Virginia through western North Carolina and extreme eastern Tennessee southward into northern Georgia and South Carolina.

HABITAT Cool, highly oxygenated streams above 300 m elevation that have loose gravel, rocks, and moderate flow gradients are preferred. Adults are more likely to inhabit riffles than slower pools because disturbed water contains more oxygen and improves respiration across the skin. Larvae remain in the interstices of rocks, gravel, and leaves on the bottom of streams well away from adults.

BEHAVIOR AND ACTIVITY Adult and juvenile shovel-nosed salamanders remain in water for most of their lives, although some adults emerge from their streams on wet nights and move a short distance into the surrounding forest where they remain under litter during the day. They can be found in running water every month of the year. Shovel-nosed salamanders are not known to migrate, and little is known about their movement activities.

FOOD AND FEEDING Insects that live in the water or fall into it are the primary prey of adults and juveniles, which also eat snails, crayfish, beetles, flies, mayflies, caddisflies, wasps and bees, and other salamanders such as two-lined salamanders that live in the same streams. Adults occasionally eat juveniles and larvae of their own species. The larvae eat a wide variety of aquatic insects. Feeding takes place at night, usually while the salamanders are hiding beneath stones and watching for prey to swim by or while they are perched on rocks.

Juvenile shovel-nosed salamanders bear pairs of light spots that disperse as they attain maturity.

REPRODUCTION The mating behavior has not been described, but presumably mating occurs in underwater riffles during late winter and early spring. Females lay clutches of 20–65 eggs in late spring and summer. Eggs are attached singly or

Body coloration consists of two rows of highly irregular yellowish brown, gray, olive, or yellowish spots over a dull, dark brown, gray, or black background.

in clusters of 2–4 on the underside of large rocks in flowing water. Hatching occurs from mid-August through September. Larvae may remain in the larval stage for as long as 3 years before absorbing their gills and going through metamorphosis.

PREDATORS AND DEFENSE Predators of shovel-nosed salamanders include several species of carnivorous fish such as sculpin and brook trout, northern watersnakes, and large stonefly nymphs. Some individuals practice cannibalism. Fleeing for the shelter of stones and leaves on the stream bottom is the primary means of escaping predation.

CONSERVATION The high-elevation streams used by these salamanders are susceptible to siltation from upslope sources such as road building, forest clearing, and home construction. Their need for highly oxygenated streams makes them sensitive to pollution of all sorts and to siltation from upstream activities and dam construction. Like other stream salamanders, shovel-nosed salamanders are collected and used for fish bait, but less commonly now than a few decades ago.

TAXONOMY The shovel-nosed salamander was described in 1899 and remained in its own genus (*Leurognathus*) for many years until genetic evidence indicated that this species is closely related to the *Desmognathus* salamander group. Its genetic relationship with the black-bellied salamander is unclear. High levels of variation in color and pattern across the range and in individual populations suggest that there may be hidden species within the shovel-nosed salamander group, and the taxonomy may be revised.

Shovel-nosed salamanders remain in water in southern Appalachian mountain streams for most of their lives.

Adult many-lined salamanders are distinctive in having a dark brown band that runs through the eye onto the back and black streaking on the body.

How do you identify a many-lined salamander?

BODY SHAPE
Slender with a short tail

NUMBER OF COSTAL GROOVES
18

BODY COLOR
Brown and yellow

PATTERN
Broad line along back and numerous thin lines on sides

BODY SIZE
Adults to 4.5 inches in total length

Many-lined Salamander   *Stereochilus marginatus*

DESCRIPTION The adult is a small, slender salamander with a relatively short tail. A dark brown band on the back runs through the eye and onto the tail. The sides have a series of fine, black longitudinal lines or streaks that may appear as a series of spots. The belly is yellow with black flecks. The tail is flattened side to side toward the tip and has a keel.

LARVAE AND JUVENILES The larvae and juveniles are distinctly dark brown above and yellow below. The head and body may have small yellow or white spots. Older larvae may be mottled on the back.

SIMILAR SPECIES Dwarf salamanders are smaller and have a broad brown stripe down the back bordered by black lines, a long tail, no thin lines on the sides, and

Large many-lined salamander larvae (top) have dark backs and yellow bellies. Small larvae (bottom) may be reddish and have distinctive spots on the body and tail.

only four toes on the rear feet. Mud salamander larvae are light brown with a few black spots rather than dark brown with yellow spots.

DISTRIBUTION Many-lined salamanders are found in the Coastal Plain from southeastern Virginia to northern Florida.

HABITAT These salamanders occupy gum and cypress swamps, woodland ponds, ditches, canals, sluggish streams, and other shallow permanent wetlands. The mucky pools, sphagnum mats, and leaf litter often found in these wetlands are used extensively for concealment. All life stages occupy shallow, acidic waters.

BEHAVIOR AND ACTIVITY Little is known about the seasonal activity patterns and movements of this secretive salamander. Individuals spend their entire lives in small patches of wetlands, aestivating in small cavities in the substrate when the water dries up and sometimes hiding under logs and other surface objects adjacent to wetlands.

Many-lined salamanders inhabit swamps, ditches, and other wetlands containing sphagnum and other dense vegetation.

FOOD AND FEEDING The available information indicates that many-lined salamanders eat a diverse selection of prey, including small clams, amphipods, isopods, and insects and their larvae.

REPRODUCTION Reproduction takes place in water in the fall. Courtship and mating behaviors are incompletely described but are probably similar to those of other members of the plethodontid group; males produce spermatophores that are picked up by females for internal fertilization. Females lay 16–121 eggs in water or on nearby land in the winter, attaching them singly to the underside of logs, debris, or other objects, or depositing them in aquatic mosses in shallow water. Females do not attend their eggs. Incubation lasts 1–2 months, and eggs hatch in late March and April.

PREDATORS AND DEFENSE Natural predators have not been documented but likely include predatory aquatic insects, snakes, fish, and wading birds. Defense mechanisms or behaviors are unknown.

many-lined salamander
Stereochilus marginatus

Loss of wetlands is a major threat to the many-lined salamander.

CONSERVATION The primary threat to this salamander is loss of wetlands to drainage and conversion to urban, agricultural, and industrial uses. It is a Species of Special Concern in Virginia because its distribution is limited and the wetland habitats it requires are being destroyed.

TAXONOMY The many-lined salamander was first described in 1856. No subspecies are recognized.

Waterdogs

Black Warrior River Waterdog	*Necturus alabamensis*
Gulf Coast Waterdog	*Necturus beyeri*
Neuse River Waterdog	*Necturus lewisi*

DESCRIPTION Adult waterdogs are large, robust salamanders with four toes on each of their four feet; large, feathery gills; and a large tail that is flattened side to side. Three species of similar appearance occur in the Southeast. The Black Warrior River waterdog is reddish brown to nearly black with minimal black spots on the back and a uniformly gray belly. The Gulf Coast waterdog has a brown body with numerous irregular, brown to black spots on the head, body, tail, and belly, although the center of the belly is unmarked. The Neuse River waterdog has a rusty brown body with numerous large, bluish black spots or blotches on the body and tail, and a dull brown to gray belly with spots.

Adult waterdogs have large, feathery gills.

How do you identify a waterdog?

BODY SHAPE
Robust; big, bushy gills; flattened tail

NUMBER OF COSTAL GROOVES
None

BODY COLOR
Gray, brown, to black

PATTERN
Irregular black spots

BODY SIZE
Adults to 11 inches in total length

Black Warrior River waterdog larva (top) and Gulf Coast waterdog larvae (bottom)

LARVAE AND JUVENILES Larval Black Warrior River waterdogs have three dark stripes separated by two cream or yellow dorsal stripes. Larval Gulf Coast waterdogs have a few light dorsal spots. Neuse River waterdog larvae are uniformly brown with a white spot behind each eye. Juvenile Black Warrior River waterdogs are brownish and have remnants of the dark larval stripes. Gulf Coast waterdog juveniles have yellow spots on the head and body and along the edge of the tail fin. Immature Neuse River waterdogs have spots on a gray background; small ones have a light stripe down the back and dark sides with black flecking.

SIMILAR SPECIES Dwarf waterdogs usually have no markings on a dark gray body and a white line down the middle of the belly. Mudpuppies have a conspicuous dark line through each eye and irregular black spots on a gray body and tail. All of the mole salamander (*Ambystoma*) larvae have feathery gills, but they are small, do not occur in rivers, and have five toes on each hind foot.

DISTRIBUTION Black Warrior River waterdogs are found only in the northern half of Alabama in the Black Warrior River watershed. Gulf Coast waterdogs are in central Alabama, all of Mississippi, eastern Louisiana, and eastern Texas. Neuse River waterdogs occupy the Neuse River watershed in North Carolina. See Taxonomy for additional comments.

HABITAT Logjams, detritus, and leaf litter are important habitat features of the streams and rivers that adults and juveniles of all three species occupy, although Neuse River waterdogs do occur in streams with a clay or hard soil substrate. Black Warrior River waterdogs require large streams with moderate flows and alternating pools and riffles. Hatchlings and juveniles use rocks and other objects for shelter; waterdogs of all ages use burrows of crayfish and other animals in underwater stream banks.

BEHAVIOR AND ACTIVITY In the Deep South, waterdogs are inactive during hot summer periods when invertebrate prey are scarce, and are most active during colder months. Neuse River waterdogs are more active in spring and fall than in winter and summer. All three species are more ac-

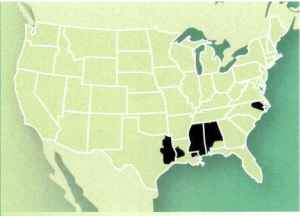

- Black Warrior River waterdog
  Necturus alabamensis
- Gulf Coast waterdog
  Necturus beyeri
- both Black Warrior River and Gulf Coast waterdogs
- Neuse River waterdog
  Necturus lewisi

Black Warrior River waterdogs are brownish with a few black spots on the back and a uniformly gray belly. All waterdogs have feathery gills and a large, side-to-side-flattened tail.

All waterdogs have four toes on each foot, as seen in this Gulf Coast waterdog from Wakulla County, Florida.

Neuse River waterdogs have large bluish black blotches on the brown body and tail.

Neuse River waterdogs occur only in the Neuse River watershed in North Carolina.

tive after moderate rains but cease activity during heavy rainfall. They forage at night and remain hidden in leaf litter and under objects on the bottom during the day. Adults in burrows carry gravel and sand to the burrow opening, presumably to aid in concealment. Both sexes actively defend their burrows.

FOOD AND FEEDING Adults and juveniles eat almost any animals they can capture and swallow, including many types of insects and their larvae, clams, snails, worms, slugs, leeches, spiders, crayfish, and small fish. A Neuse River waterdog was documented to have eaten a worm snake.

REPRODUCTION Adults mate in water from late fall through winter. Females attach about 4–40 eggs in oblong clusters to the underside of logs, rocks, and other objects on the stream bottom in April and May. Males may stay with the eggs until they hatch in June, July, and August.

PREDATORS AND DEFENSE Northern watersnakes are known predators of waterdogs, and fish and crayfish are probably common predators as well, although no direct natural observations have been reported. Their inactivity in warm months may be a way to hide from potential aquatic predators. Skin secretions may deter some predators.

CONSERVATION Waterdog populations are threatened by stream pollution, siltation, channelization, and clearing of logjams and debris. The Black Warrior River waterdog is listed in Alabama as a Species of High Conservation Concern and is being considered for protection under the U.S. Endangered Species Act. Almost nothing has been published about the life history and ecology of this rare species, so the extent of its danger is unknown. The Neuse River waterdog is a Species of Special Concern in North Carolina.

TAXONOMY The Black Warrior River waterdog was described in 1937, the Gulf Coast waterdog in 1937, and the Neuse River waterdog in 1924. No subspecies are currently recognized for these species, which all go by the common name mudpuppy. Several additional species will likely be described in the future from Georgia, northern Florida, and Alabama.

This dwarf waterdog from Crawford County, Georgia, has the large gills and drab, un-marked body typical of the species.

Dwarf Waterdog

Necturus punctatus

DESCRIPTION Dwarf waterdogs are the smallest and most slender of the waterdog group. The body is usually a uniform slate gray to brown without spots, although populations in the Cape Fear and Lumber river drainages in North Carolina have some irregular dark spotting on the tail. Each of the four feet has four toes. The large, red, bushy gills are conspicuous. The midline of the belly is pale cream to white without spots.

LARVAE AND JUVENILES The larvae are brown but otherwise resemble adults. Hatchlings and juveniles are uniformly brown without spots or stripes. The tail fin may be mottled, and the belly is bluish white.

The dwarf waterdog has conspicuously large, red, bushy gills.

How do you identify a dwarf waterdog?

BODY SHAPE
Slender; bushy gills; flattened tail

NUMBER OF COSTAL GROOVES
None

BODY COLOR
Brown to gray

PATTERN
Uniformly colored, without spots

BODY SIZE
Metamorphs 2 inches; adults to 7.5 inches in total length

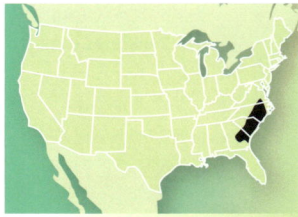

● dwarf waterdog
  Necturus punctatus

Juvenile dwarf waterdogs are uniformly brown but may have dark mottling on their tails.

SIMILAR SPECIES All other waterdogs and the mud-puppy are larger, more robust, and have irregular spots as adults. Juvenile dwarf sirens may have longitudinal light and dark stripes. Neuse River waterdogs have a rusty brown body, dark spots or blotches on the body and tail, and a dull brown to gray belly with spots.

DISTRIBUTION Dwarf waterdogs occur in the Coastal Plain and lower Piedmont from southern Virginia through the Carolinas to central Georgia. They are rare to absent from the mainstream Neuse and Tar rivers in North Carolina.

HABITAT Sluggish streams with an accumulation of silt, debris, and leaves appear to be preferred, although this may reflect modern impacts to the streams because dwarf waterdogs can also be abundant in clear-flowing blackwater streams. Adults and juveniles are most often found in leaf beds in shallow water.

BEHAVIOR AND ACTIVITY Winter appears to be the primary season of activity, but dwarf waterdogs can be found at all times of the year. Individuals may aggregate in leaf beds in the winter. Adults, larvae, and juveniles are apparently nocturnal, and all life stages are aquatic.

FOOD AND FEEDING Adults and juveniles are generalist predators that will eat any animal they can fit into their mouth. Worms, insects and their larvae, and aquatic invertebrates have been identified as prey.

REPRODUCTION Courtship and mating have not been described but probably resemble the behaviors observed in other waterdogs. Mating probably occurs in winter, with egg deposition in March to May. Females lay 15–55 eggs, presumably under objects in the stream as other waterdogs do, although no nests have been described. Larvae have been caught in April and November, but the length of the larval period is unknown.

PREDATORS AND DEFENSE Predators presumably include some fish, otters, wading birds, sirens, and watersnakes and may also include other dwarf waterdogs. Dwarf waterdogs are not known to produce noxious skin secretions.

CONSERVATION Although the species is apparently not common anywhere, dwarf waterdogs are not protected in any state in which they occur. Stream pollution and siltation from agricultural, residential, and industrial sources undoubtedly affect this gill-breathing salamander. The conservation status needs modern evaluation throughout the species' range.

TAXONOMY The dwarf waterdog was first described in 1850. No subspecies are recognized.

Some dwarf waterdog populations in the Cape Fear and Lumber river drainages in North Carolina have some irregular dark spotting on the tail.

Mudpuppies are the largest of the waterdog group and occur primarily in the central and northern United States.

BODY SHAPE
Robust; bushy gills; flattened tail

NUMBER OF COSTAL GROOVES
None

BODY COLOR
Gray to dark brown

PATTERN
Irregular spots or blotches on body and tail

BODY SIZE
Adults to 19 inches in total length

Mudpuppy

Necturus maculosus

DESCRIPTION Mudpuppies are large aquatic salamanders with large, dark red, feathery gills; four limbs with four toes on each foot; and a large, flattened, paddle-like tail. Adults are usually rusty brown or occasionally grayish, with irregular bluish black spots or blotches on the body and tail. The belly is also gray with black spots. A dark line passes from the nostril on the flattened snout to the gills on each side of the head. Individuals of the Red River mudpuppy subspecies have a dark stripe down the back bordered on each side by a faint light stripe and large spots or blotches on the body. The belly is light gray with no spots.

LARVAE AND JUVENILES Larvae have a broad, dark line down the middle of the back that is bordered on each side by a yellow stripe. Juveniles have the same pattern, but the yellow and black stripes on the sides are more pronounced.

SIMILAR SPECIES The only waterdog that may occur in the same geographic area is the Black Warrior River waterdog, which is reddish brown to nearly black and usually has no black spots on the back and a uniformly gray belly. All other waterdogs occur well outside the known range of the mudpuppy. The hellbender has a broad, flat body with wrinkled folds of skin.

● mudpuppy
Necturus maculosus

The dark stripe passing through the eye (left) is pronounced on most individuals. Gill size may be reduced in highly oxygenated water and in captivity.

Mudpuppies (right) have large, dark red, feathery gills, especially in ponds and lakes.

Larvae and juveniles have a broad, dark line down the middle of the back that is bordered on each side by a yellow stripe.

DISTRIBUTION In the Southeast, mudpuppies occur in Kentucky, south-western Virginia, Tennessee, western North Carolina, northern Alabama and Georgia, Mississippi, and northern Louisiana. The Red River mud-puppy occurs from north-central Louisiana west of the Mississippi River to southeastern Kansas.

HABITAT Adults occupy clear to silted lakes, reservoirs, ditches, streams, and rivers in the Southeast. Flat rocks, logs, debris, crayfish burrows, un-dercut banks, and rock slabs provide hiding places. Water with high oxygen content below riffles is preferred. Juveniles inhabit pools with a substrate of litter and organic debris.

BEHAVIOR AND ACTIVITY Mudpuppies are active in all months of the year and probably remain in the same vicinity for most of their long lives (to at least 34 years). Individuals occasionally move upstream and into tribu-taries.

FOOD AND FEEDING Mudpuppies eat almost any creature they can get into their mouth; only the size of their gape limits what they can consume.

Pollution and siltation from agricultural, residential, and industrial sources threaten streams and mudpuppy populations.

Crayfish and small fish are important components of the diet, but a wide variety of insects and their larvae, worms, snails, some amphibians such as dusky and two-lined salamanders, small turtles, fish eggs, and eggs of other mudpuppies are also eaten.

REPRODUCTION Mating probably occurs from fall through early spring in most of the Southeast. Males search for females in shallow-water retreats. The male courts a female to entice her to pick up a sperm packet that he has deposited on the substrate. Females turn upside down to attach each of their 20–200 eggs singly to the roof of a nesting chamber excavated beneath a rock, log, or other structure embedded on the stream bottom at a depth of less than 2 feet. Hatching occurs 1.5–2 months later in early summer, but the hatchlings remain in the chamber until the ex-ternal yolk sac is absorbed, as long as 6–8 weeks. Emergence from the nest chamber usually occurs in August. Females often guard the eggs until they hatch.

PREDATORS AND DEFENSE Animals that eat mudpuppies include crayfish, predatory fish, carnivorous turtles, watersnakes, herons, otters, and larger mudpuppies. The larvae and smaller individuals may also become food for predatory larvae of aquatic insects. Defense is usually a quick escape with a rapid movement of the flattened tail.

CONSERVATION Pollution and siltation from agricultural, residential, and industrial sources threaten streams and their salamander populations throughout the range of this species. The mudpuppy is a Species of Special Concern in North Carolina. The ecology and life history have been well studied in the northern portion of the range, particularly in the Great Lakes area, but little is known about this species in the Southeast.

TAXONOMY Two recognized subspecies are found in the Southeast. The common mudpuppy (*Necturus m. maculosus*), described in 1818, is found from Kentucky and southwestern Virginia south to northern Georgia, Alabama, and Mississippi. The Red River mudpuppy (*N. m. louisianensis*), described in 1938, is in northern Louisiana.

This mudpuppy from Kentucky has the dark body spots characteristic of the species and a dark line from the snout to the gills.

How do you identify a dwarf siren?

BODY SHAPE
Slender; tiny front legs, each bearing three toes, and no rear legs; bushy gills

NUMBER OF COSTAL GROOVES
29–37

BODY COLOR
Brown to black

PATTERN
Yellow stripes on back or sides

BODY SIZE
Adults to 9 inches in total length

Dwarf Sirens

Northern Dwarf Siren   *Pseudobranchus striatus*

Southern Dwarf Siren   *Pseudobranchus axanthus*

DESCRIPTION Dwarf sirens are small, eel-like salamanders with two legs and external gills. Each of the two front legs has three toes. Northern dwarf sirens are brownish to black with a yellow line on each side that runs from the snout to the tail. Many individuals lack sharply defined dorsal stripes, although they may have thin dark lines on the back. The Gulf hammock dwarf siren subspecies (*P. s. lustricolus*) has three narrow yellow stripes on the back and two on each side; the slender dwarf siren (*P. s. spheniscus*) has no stripes on the back and two stripes on each side. Southern dwarf sirens are light gray to brownish black with parallel yellow or tan stripes on the back and sides that run from the head to the tail tip. Stripe color and number vary among populations. Narrow-striped dwarf sirens (*P. a. axanthus*) have five stripes that are weakly defined or distinct but have irregular margins; the Everglades dwarf siren (*P. a. belli*) has five distinct stripes.

LARVAE AND JUVENILES Hatchlings of both species have a fin on the back extending from the head to the tip of the tail. Hatchling northern dwarf sirens look like adults but are blacker and have an overall reddish hue. Southern dwarf siren hatchlings are brown with a yellowish stripe down the back and on each side.

Northern dwarf sirens have a dark body with a bold yellow line on each side that runs from the snout along the body and tail.

The Gulf hammock dwarf siren subspecies (left) has three narrow yellow stripes on the back and two on each side.

Narrow-striped dwarf sirens (right) have five stripes that are weakly defined or distinct but have irregular margins.

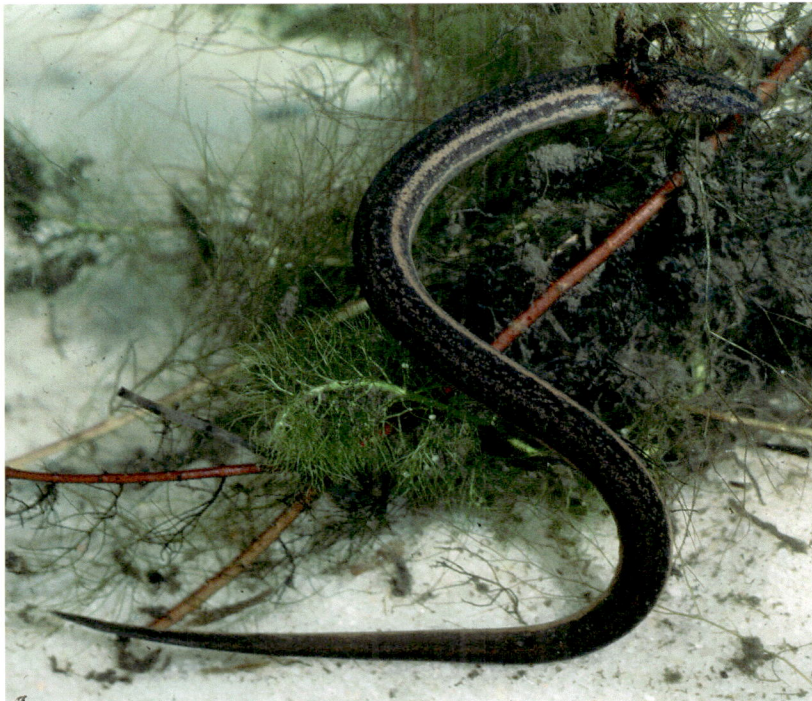

The slender dwarf siren has no stripes on the back and two stripes on each side.

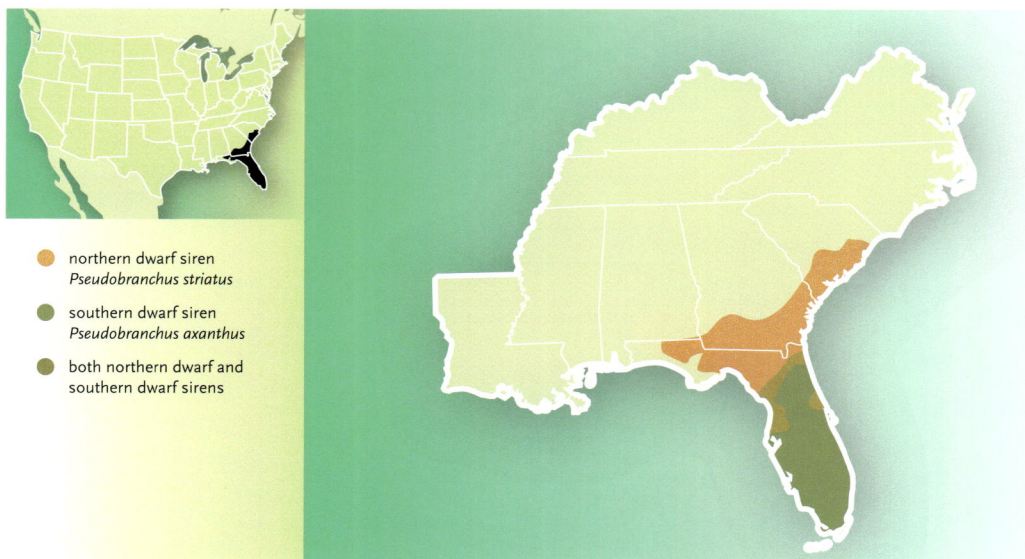

- ● northern dwarf siren
  Pseudobranchus striatus
- ● southern dwarf siren
  Pseudobranchus axanthus
- ● both northern dwarf and
  southern dwarf sirens

SIMILAR SPECIES The greater siren is similar in appearance but much larger (to 38 inches) and has 36–40 costal grooves and four toes on each foot; juveniles have a light stripe along the side of the body. The lesser siren (to 27 inches) has 31–35 costal grooves, and juveniles have red or yellow lines on the head. Amphiumas have two pairs of tiny limbs and lack external gills.

DISTRIBUTION Northern dwarf sirens are found from the lower Coastal Plain of South Carolina south through Florida to Volusia County on the east and Hernando County on the west and in much of the Florida panhandle. Southern dwarf sirens occur only in peninsular Florida.

HABITAT Dwarf sirens occupy cypress and gum ponds, swamps, ditches, marshes, limestone sinks, and other vegetation-choked freshwater wetlands. They prefer thick vegetation in the water column and decaying vegetation and mucky soils along the wetland margin. Mats of the exotic plant water hyacinth are frequently used microhabitats. Northern dwarf sirens are more often associated with acidic cypress swamps. Southern dwarf sirens are more often associated with freshwater marshes and prairie ponds.

BEHAVIOR AND ACTIVITY Dwarf sirens form cocoons in the mud when their wetlands begin to dry and can aestivate for several months encapsulated underground. Like other aestivating sirens, they absorb their gills and undergo changes in body mass and physiology during this time. They occasionally aestivate in S-shaped burrows as much as a foot below the surface of the substrate. Dwarf sirens occasionally vocalize with a faint, high-pitched squeak.

FOOD AND FEEDING Dwarf sirens eat a wide array of small aquatic invertebrates, including many insects and their larvae, amphipods, ostracods, and worms. Their small mouth size limits the size and therefore the range of prey these salamanders can consume.

REPRODUCTION Courtship and mating have not been described. Fertilization is likely external because males lack the ability to produce spermatophores. Females lay small numbers (probably fewer than 20) of eggs singly or in small clusters attached to vegetation in November to March. Clutch size is unknown. Eggs probably hatch in a few weeks, but individuals may take as long as 2 years to reach maturity.

PREDATORS AND DEFENSE Known predators include striped crayfish snakes, mud snakes, and southern watersnakes. Small alligators, turtles, wading birds, fish, and some predatory invertebrates undoubtedly eat dwarf sirens as well. Some adults emit an audible yelp when captured or prodded, a sound that might deter some predators. Their most effective defense if captured is to thrash about, because they are very slippery.

CONSERVATION Too little is known about the population status of dwarf sirens anywhere to determine if numbers or populations have been lost or are declining. Undoubtedly, complete loss of wetlands and degradation of wetlands resulting from agricultural, residential, and silvicultural activities have caused population losses. The Gulf hammock dwarf siren has not been captured in Citrus and Levy counties, Florida, since 1951 and may be extinct.

TAXONOMY Three subspecies are recognized for the northern dwarf siren (*Pseudobranchus striatus*). The broad-striped dwarf siren (*P. s. striatus*) was described in 1824; the Gulf hammock dwarf siren (*P. s. lustricolus*) was described in 1951; and the slender dwarf siren (*P. s. spheniscus*) was described in 1949. The broad-striped dwarf siren is found from southern South Carolina south to the northeastern coastal flatwoods of Florida. The Gulf hammock dwarf siren is limited to Citrus and Levy counties, Florida. The slender dwarf siren occurs from northwestern peninsular Florida to south-central Georgia and in much of the Florida panhandle. The southern dwarf siren (*Pseudobranchus axanthus*), described in 1942, has two recognized subspecies: the narrow-striped dwarf siren (*P. a. axanthus*) and the Everglades dwarf siren (*P. a. belli*), described in 1952. The narrow-striped dwarf siren occurs in the northern half of Florida, and the Everglades dwarf siren is found in the southern half of the state.

Did you know?

Using backhoes to clean out lake basins in the Southeast unearths and kills thousands of giant salamanders (amphiumas and sirens) aestivating in the bottom mud. Drawing down reservoirs and scraping the bottom for maintenance purposes can likewise have severe consequences for local salamanders.

Lesser sirens can be more than 2 feet long. The gills are visible throughout the adult's life. Most individuals are grayish all over.

BODY SHAPE
Eel-like and slender; feathery external gills; one pair of legs

NUMBER OF COSTAL GROOVES
31–35

BODY COLOR
Dark gray to black

PATTERN
Mostly uniform, but some have tiny black dots

BODY SIZE
Adults to 27 inches in total length

Lesser Siren

Siren intermedia

DESCRIPTION Lesser sirens are eel-like in body form and have external gills and only one pair of small front legs. Each of the two limbs has four toes. Adults are uniformly gray to olive green to nearly bluish black. Some individuals have tiny black dots on the back and sides. The belly is uniformly light gray. There are 31–35 costal grooves along the body between the front legs and the cloaca.

LARVAE AND JUVENILES Hatchlings have partially developed limbs and a dorsal fin that runs from the head to the tip of the tail. Hatchlings and small juveniles have a red band across the snout and along the side of the head but are otherwise uniformly brown to gray.

SIMILAR SPECIES Greater sirens are similar in appearance but are much larger as adults (to 38 inches) and have 36–40 costal grooves. Juvenile greater sirens have a light stripe along the side of the body and no red markings on the head. Dwarf sirens are small with several stripes and three toes on each limb. Amphiumas do not have external gills.

DISTRIBUTION Lesser sirens occur in the Southeast from the Coastal Plain of Virginia south to southern Florida, west through central Georgia and

lesser siren
Siren intermedia

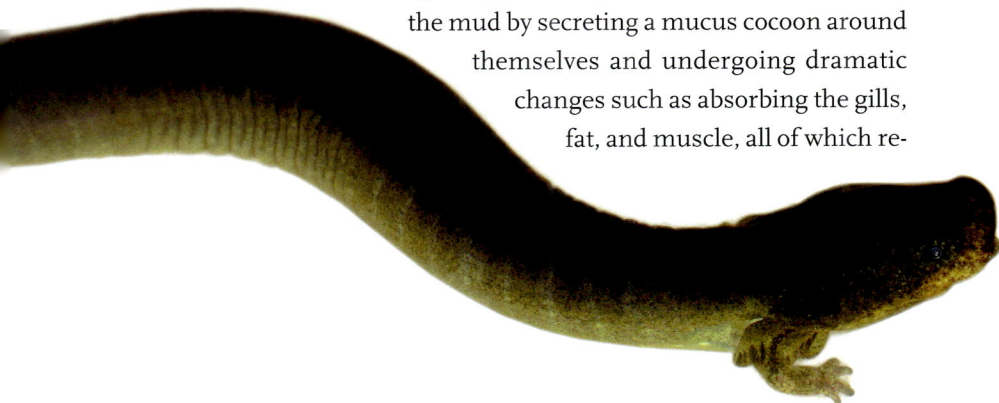

Lesser sirens have external gills and only one pair of small front legs.

Alabama through Louisiana, and north to western Tennessee and Kentucky. Eastern lesser sirens are found from Virginia through Alabama, and western lesser sirens occur in states west of Alabama and north to western Kentucky.

HABITAT Lesser sirens live in ditches, ponds, sloughs, backwaters of small sluggish streams, ponds, shallow lakes, marshes, and swamps where aquatic vegetation and organic sediment are plentiful. Hatchlings and juveniles prefer vegetation-choked habitats and root masses. Lesser sirens can withstand water low in oxygen and acidity, and can aestivate in the mud when their wetlands dry.

BEHAVIOR AND ACTIVITY Lesser sirens have both gills and lungs. They occasionally venture beyond the water's edge to hide under objects adjacent to water, and they may move considerable distances over land when rains create temporary waterways. Individuals aestivate in the mud by secreting a mucus cocoon around themselves and undergoing dramatic changes such as absorbing the gills, fat, and muscle, all of which regenerate when the pond refills. Activity slows in winter months, and lesser sirens in northern latitudes may hibernate in aggregations underground. Except when migrating to a nest or dispersing over land, these sirens remain within a small aquatic range less than 20 feet square. They feed at night. Lesser sirens sometimes make a clicking or yelping noise that could be a communication with other sirens.

FOOD AND FEEDING Juveniles and adults eat a diverse array of prey, including insects and their larvae, worms, snails, freshwater clams, tadpoles, larval salamanders, and small fish. Egg cannibalism has been reported.

REPRODUCTION Mating has not been described. Females lay 200–500 eggs in February and March in root masses and other dense vegetation. Larvae hatch in 1.5–2.5 months and remain in dense vegetation for protection from predators.

Some individuals have tiny black dots on the back and sides.

PREDATORS AND DEFENSE Alligators, southern banded watersnakes, cottonmouths, mud snakes, largemouth bass, and wading birds are known predators. Feral hogs likely kill many sirens during droughts while the sirens are in cocoons in the mud. The tiny larvae and young sirens may be eaten by predaceous diving beetles, dragonfly nymphs, and other predatory insects. Nocturnal feeding may be a predator-avoidance strategy. One researcher noted shrill distress vocalizations.

CONSERVATION Lesser sirens are apparently abundant in many areas, but populations suffer from aquatic pollution and habitat loss and fragmentation resulting from flood control programs and development. Feral hogs that root up aestivating salamanders in drying wetlands probably kill large numbers.

TAXONOMY Two subspecies, both described in 1826, are recognized: the eastern lesser siren (*Siren i. intermedia*, with 31–33 costal grooves) and the western lesser siren (*S. i. nettingi*, with 35 costal grooves). There are several yet-to-be-described species in the Southeast. Presumably all possess the same basic life histories and ecology described here.

Adults are uniformly gray to olive green to nearly bluish black.

Adult greater sirens can be more than 3 feet long. Some individuals such as this one have gold flecking. Sirens have two front legs but no hind legs.

How do you identify a greater siren?

Greater Siren

Siren lacertina

DESCRIPTION Greater sirens are large, eel-shaped salamanders with a pair of large, bushy gills; a shovel-shaped head; and one pair of front legs with four toes on each limb. The body is olive, dark brown, or nearly black with a gray belly. Individuals in some locations have well-defined gold flecking on the head and body. The lower sides are lighter than the back and often have faint greenish or yellowish dots or dashes. Some greater sirens are mottled with tiny yellowish dots on a gray or brown body. The belly has small greenish or yellowish flecks.

LARVAE AND JUVENILES The larvae have fins on the back and tail, and both larvae and young juveniles are brown to gray with a yellowish stripe along each side from the gills to the base of the tail. The stripe fades within a year.

SIMILAR SPECIES Lesser sirens are similar in appearance but are smaller as adults (to 27 inches) and have fewer costal grooves (31–35 rather than 36–40). Juvenile lesser sirens lack

BODY SHAPE
Eel-like shape; large, bushy gills; one pair of legs

NUMBER OF COSTAL GROOVES
36–40

BODY COLOR
Olive, gray, or black

PATTERN
None or with small gold spots and flecking

BODY SIZE
Adults to 38 inches in total length

Larvae and young juveniles are brown to gray with a yellowish stripe along each side from the gills to the base of the tail.

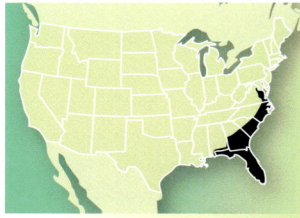

● greater siren
Siren lacertina

Greater sirens are large and robust. Their larger size and 36–40 costal grooves distinguish them from lesser sirens.

the light stripe along the sides of the body. Dwarf sirens are small with stripes and three toes on each limb. Amphiumas have four tiny limbs and lack external gills.

DISTRIBUTION Greater sirens occur from southern Maryland in the Coastal Plain south through peninsular Florida and west to southern Alabama.

HABITAT Adults and juveniles occupy semipermanent to permanent aquatic habitats with abundant vegetation, such as streams, rivers, ditches, canals, freshwater marshes, Carolina bays, farm ponds, and lakes. Juveniles prefer shallow water but move into deeper water as they mature. Sirens use their shovel-shaped head to burrow in the bottom of rivers and wetlands.

BEHAVIOR AND ACTIVITY Greater sirens are aquatic salamanders with functional gills. They may move short distances over land between bodies of water during rains, but probably must be submerged to move more than a few feet. Individuals may be active in all months of the year, but activity peaks during the early spring and summer months. Adults and juveniles are active at night and hide in burrows and under debris and litter on the bottom or in thick vegetation during the day. When wetlands dry, sirens secrete mucus that forms a cocoon around their body and aestivate until water returns, subsisting in the meantime on their own body tissues. Laboratory studies indicate that large greater sirens can aestivate for more than a year under ideal conditions. Greater sirens can make clicking sounds when in proximity to others, but their purpose is unknown.

FOOD AND FEEDING Greater sirens are generalists that feed on a wide variety of prey, particularly freshwater clams, snails, crayfish, and many types of insects and spiders; they also scavenge. They may be able to digest plant matter as well, because individuals are often found with large amounts of algae in their stomach. Adults occasionally eat aquatic vertebrates such as pond-breeding salamanders and small fish.

REPRODUCTION Males apparently fertilize eggs as the females lay them, because no spermatophores are produced. Little is known about courtship and mating. Females lay 100–500 eggs singly or in small groups attached to leaf litter in shallow water in February, March, and April. Hatching occurs after an incubation period of about 2 months. Females may remain with their eggs and guard them from predators.

Greater sirens are slippery and hard to hold. This photo captures a moment when the three individual sirens shown were not moving.

PREDATORS AND DEFENSE Alligators, mud snakes, southern banded watersnakes, eastern green watersnakes, and wading birds are known predators. Feral hogs probably dig up sirens and eat them, especially during droughts when the salamanders are aestivating in the mud. Predatory insects and fishing spiders probably eat larvae and small juveniles. Several noises made by greater sirens have been suggested to deter some predators, especially birds and mammals, but confirmation that such soft sounds are effective is lacking. Sirens' primary defense is mucus secreted by skin cells that makes their body slippery and difficult to hold onto as they thrash and swim away.

CONSERVATION Greater sirens appear to be relatively common in many parts of their range except in peripheral areas. The species is listed as Endangered in Maryland. Stream and river pollution may harm individuals and populations living in secondary waters. The effects of feral hog consumption of aestivating sirens may be severe. The conservation status of this aquatic species needs evaluation in all southeastern states.

TAXONOMY The greater siren was first described in 1766. No subspecies are currently recognized, although one additional species from Alabama and northwestern Florida may be added to the greater siren group in the near future.

AQUATIC CAVE SALAMANDERS

The Georgia blind salamander has a large, flat head; squared-off snout; long external gills; and long spindly legs. The tiny eyes are dark spots, and the body is translucent.

Georgia Blind Salamander *Haideotriton wallacei*

DESCRIPTION These unusual little salamanders have a large, slightly flattened head; a squared-off snout; long, bright red external gills; four long, spindly legs; and a flattened tail with fins. Adults retain all of the typical larval characteristics. The degenerate eyes are dark spots buried below the skin. The body is pink to silvery white and translucent, and may have traces of dark specks or blotches.

LARVAE AND JUVENILES The larvae have not been described because they have never been seen by scientists. Juveniles resemble adults in most characters. Small individuals have some dark pigment. They do not metamorphose.

SIMILAR SPECIES The other troglobitic salamanders with which this species might be confused, such as the Tennessee cave salamander and grotto salamander, occur well outside the range of the Georgia blind salamander.

DISTRIBUTION Georgia blind salamanders are found in the Coastal Plain of southwestern Georgia and in several counties in the Florida panhandle. The only known localities are in caves that provide access to the Floridan Aquifer.

How do you identify a Georgia blind salamander?

BODY SHAPE
Slender; long head; tail flattened from side to side

NUMBER OF COSTAL GROOVES
11–13

BODY COLOR
Pink and white

PATTERN
None

BODY SIZE
Adults to 3 inches in total length

Young Georgia blind salamanders (top) have more dark pigmentation than adults (bottom).

HABITAT People have found Georgia blind salamanders in limestone caves, sinkholes, and artesian springs, some with deep silt layers. These completely aquatic salamanders occupy subterranean streams and pools, especially in areas where bat guano provides nutrients to the aquatic system.

BEHAVIOR AND ACTIVITY Georgia blind salamanders are very sensitive to light and movements in the water. They will swim quickly in an upward spiral when disturbed but then settle back down on the substrate and remain motionless. Their seasonal activity period is unknown but may be regulated by stream flows and the timing of nutrient input into the caves.

FOOD AND FEEDING Georgia blind salamanders either bump into potential prey items or sense their presence with the lateral line system, a row of sensory cells on the sides that detect vibrations in the water. Adults and juveniles eat small isopods, amphipods, copepods, mites, and small beetles.

REPRODUCTION These small salamanders reach maturity at a total length of about 2 inches. Females carrying enlarged eggs have been found in May and November. Nothing is known about mating behavior, clutch size, egg-laying dates, or other aspects of reproduction.

PREDATORS AND DEFENSE Georgia blind salamanders co-occur with the Dougherty Plain blind crayfish, which may be a primary predator. Several species of fish are known from the same underground water system, and they, too, could easily eat these salamanders. Predation has never been observed in nature.

CONSERVATION This unique salamander is threatened by any human activity that affects caves and aquifers, including chemicals associated with agriculture, urbanization, and industrial activities. Loss of aquifer water from drawdowns for agriculture and drinking water for cities and towns is a potentially severe threat. The species is listed as Threatened in Georgia.

TAXONOMY The Georgia blind salamander was first known to science when specimens found in an artesian well were described in 1939. No subspecies are recognized. Some amphibian biologists place this species in the genus *Eurycea*.

Georgia blind salamander
Haideotriton wallacei

These subterranean salamanders are true paedomorphs that retain all the larval characteristics as adults. Paedomorphosis is a common trait of true troglobitic animals.

How do you identify cave salamanders?

BODY SHAPE
Slender; large, spatulate head; gills present; small eyes

NUMBER OF COSTAL GROOVES
17–19

BODY COLOR
Pink to salmon to brownish purple

PATTERN
Uniformly colored or with dark spots

BODY SIZE
Adults to 8 or 9 inches in total length

Permanent Cave Salamanders

Berry Cave Salamander   *Gyrinophilus gulolineatus*

Tennessee Cave Salamander   *Gyrinophilus palleucus*

DESCRIPTION Tennessee cave salamanders have a slender body; a large head with a broad, squared-off snout; small degenerative eyes; long external gills; and an oarlike tail. Gills are present throughout life. Body color varies among populations from pink, to flesh-colored without dark spots, to brown, to brownish purple with abundant black spots (see Taxonomy). Berry Cave salamanders are similar to Tennessee cave salamanders in body form, size, and color but have a more spatula-shaped snout, and some have a dark stripe or patch under the throat.

LARVAE AND JUVENILES Hatchlings have not been described. Juveniles resemble the adults in the same cave system.

SIMILAR SPECIES Larvae of the cave salamander (*E. lucifuga*) are generally uniformly pigmented with three rows of light spots down each side of the body, and they have a shorter snout. Spring salamanders lack external gills as adults. The gilled larvae of spring salamanders resemble the Tennessee cave salamander in body form but have larger eyes. Adult spring salamanders are salmon to pinkish with black spots and streaks; the larvae are yellowish brown to gray with black flecking. The eye diameter of Tennessee cave salamanders is less than 25 percent of the length of the snout but is more than 30 percent of the snout length in spring salamanders.

DISTRIBUTION These two salamander species are restricted to cave systems in central and eastern Tennessee, northwestern Georgia, and northern Alabama. Tennessee cave salamanders are found at the southeastern edge of the Cumberland Plateau in central Tennessee, northern Alabama, and northwestern Georgia. Berry Cave salamanders inhabit caves in the Valley and Ridge physiographic province in Knox, McMinn, Meigs, and Roane counties, Tennessee.

Juvenile Tennessee cave salamanders usually lack pigment and have small external gills.

HABITAT These obligate aquatic salamanders spend their entire lives in water inside caves, in shallow pools with rocky and sandy bottoms. They are paedomorphic and are true troglobites. Berry Cave salamanders have been found in pools with clay or mud substrates.

BEHAVIOR AND ACTIVITY Little is known about the behavior and ecology of these salamanders. Some individuals appear to be sedentary and rarely move from the pool in which they are first discovered; others have been known to move more than 400

Except for a few individuals like this one, Berry Cave salamanders retain their external gills throughout life. This individual has the species' distinctive oarlike tail and spatula-shaped snout.

feet within a stream. Winter and spring floods may wash some salamanders downstream, but individuals apparently seek cover under rocks or in crevices to avoid being washed away. Although these salamanders usually retain larval characters throughout their lives, laboratory experiments demonstrate that they can metamorphose, losing gills and tail fins and enlarging the eyes to normal size. A few naturally metamorphosed individuals have been found.

FOOD AND FEEDING Adults and larvae eat worms, amphipods, isopods, crayfish, beetles, fly larvae, stoneflies, mayflies, caddisflies, and small crustaceans. Cannibalism by larger individuals on smaller ones has also been reported.

REPRODUCTION Mating appears to occur in summer. Females may skip reproduction for 1 or more years. Eggs have not been described. Larvae have been found in December and February.

This is a typical Tennessee cave salamander adult with external gills. The oarlike tail is prominent in this individual.

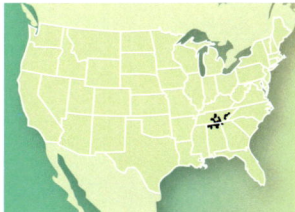

- Tennessee cave salamander
  Gyrinophilus palleucus
- Berry Cave salamander
  Gyrinophilus gulolineatus
- both Tennessee cave
  and Berry Cave salamanders

All populations of Tennessee cave salamanders are threatened by the pollution of caves and other disturbances from agriculture and urbanization.

Tennessee cave salamanders spend their entire lives in water in caves and are true troglobites.

PREDATORS AND DEFENSE Two predators have been identified: an American bullfrog inhabiting a cave, and other Tennessee cave salamanders (cannibalism). Crayfish may also eat the larvae and small adults. The salamanders readily seek refuge beneath rocks when disturbed.

CONSERVATION Cave salamanders are threatened primarily by pollution and silt entering the cave system as a result of land disturbance (runoff from agriculture, septic tanks, roads, deforestation, and urbanization). Cave systems in Tennessee are threatened as the expanding human population clears land and places houses within karst watersheds. Both species are listed as Threatened in Georgia and Tennessee and are considered Species of High Conservation Concern in Alabama. Studies of how the impacts to caves from outside sources affect these unusual salamanders will help to focus habitat management efforts that will in large part determine their fate.

TAXONOMY Two subspecies of Tennessee cave salamanders are recognized: the pinkish cave salamander (*G. palleucus palleucus*), described initially in 1954; and the spotted Big Mouth Cave salamander (*G. p. necturoides*), described in 1962. The Berry Cave salamander, described in 1965, was once considered a subspecies of the Tennessee cave salamander.

Did you know?

The eyes of salamanders that live permanently deep inside caves, such as the grotto salamander in Texas, are often no more than small, dark dots.

STREAM AND SEEP SALAMANDERS

A mature adult streamside salamander of the uniform grayish brown phase. Note the short snout and robust body.

Streamside Salamander *Ambystoma barbouri*

DESCRIPTION Most adult streamside salamanders are grayish brown to black with a dense pattern of tan to light gray markings that resemble lichens. Some adults lack the pattern on the back and are uniform in color. The belly is gray to grayish black with light grayish blotches. The snout is short, wide, and rounded.

LARVAE AND JUVENILES The aquatic larvae are uniformly dark greenish brown to olive and have feathery gills. The dorsal tail fin extends forward to behind the head. Younger larvae have three to six pairs of light blotches or saddles on the back and tail, but older larvae do not. A light-colored line of small spots may run along each side of the body. The belly is cream, and the tail fin may have greenish brown mottling. Larvae in ponds have a broad tail and large, bushy gills; larvae in streams are slim with a narrow tail and small gills. Larvae transform into terrestrial juveniles in 1–3

How do you identify a streamside salamander?

BODY SHAPE
Robust; small, rounded head

NUMBER OF COSTAL GROOVES
14 or 15

BODY COLOR
Grayish brown to black

PATTERN
Tan to gray lichen-like markings

BODY SIZE
Metamorphs 2 inches; adults to 5.5 inches in total length

The large, bushy gills and broad tail fins indicate that this streamside salamander larva is from a pond. Slender larvae with small gills are typical of stream habitats.

● streamside salamander
Ambystoma barbouri

An adult streamside salamander with a gray body and light gray lichen-like markings.

months, depending on temperature. Newly metamorphosed juveniles are uniformly brown to brownish gray with little or no pattern.

SIMILAR SPECIES The streamside salamander closely resembles the small-mouthed salamander, but geographic location can often be used to distinguish the two species. Small-mouthed salamander adults are brownish gray to black with clusters of light specks or a lichen-like pattern mostly concentrated on the sides. Juvenile small-mouthed salamanders closely resemble streamside salamander juveniles but develop the specks and lichen-like pattern within 1–2 months after metamorphosis. The teeth of the upper and lower jaws of both species occur in multiple rows; however, in streamside salamanders the inner cusp of the teeth of the upper jaw is short, rounded, and spadelike, and in small-mouthed salamanders the teeth of the upper jaw have long, pointed inner (lingual) cusps. Marbled salamander juveniles are brown to black with tiny gray to silver specks that may occur as a lichen-like pattern on the back. Spotted salamander juveniles are olive to gray with fine gray specks on the back but not on the tail. Mole salamanders are short and chunky, have a large head, and have no markings or light gray flecking. Juvenile Jefferson salamanders are uniformly dark gray, occasionally with brownish yellow specks on the sides.

A subadult streamside salamander with profuse light gray, lichen-like markings. The markings fade with age, and mature adults are uniformly colored with few markings.

DISTRIBUTION Streamside salamanders are found in north-central Kentucky, southwestern Ohio, and southeastern Indiana, with isolated populations in western Kentucky, central Tennessee, and southwestern West Virginia.

HABITAT Outside the breeding season, adults remain underground in hardwood forests adjacent to streams and ponds. Juveniles occupy similar habitat as well as upland forested areas. Breeding sites include small, fish-free headwater streams with alternating pools and riffles with large, flat limestone rocks, and a wide variety of woodland pools and pondlike wetlands.

BEHAVIOR AND ACTIVITY Adults emerge on rainy nights in late October through early March and migrate to streams and pools for mating. The females lay eggs in the water between early December and early April, then migrate back to their upland burrows. Eggs hatch from mid-March through May, and larvae remain in the aquatic habitat until they transform into juveniles 1–3 months later.

FOOD AND FEEDING The aquatic larvae eat small invertebrates and zooplankton, especially aquatic isopods. Larval densities are sometimes so high that the young salamanders eat most of the prey in their pool. Can-

Did you know?

All salamanders lay eggs, with the exception of some of the fire salamanders of Europe that are livebearers.

Streamside Salamander • 99

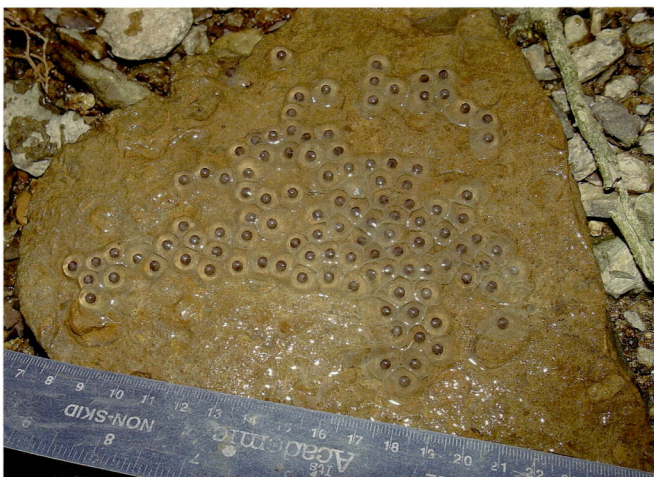

Female streamside salamanders attach their eggs in single rows under flat rocks in streams.

nibalism occurs occasionally. Both juveniles and adults are opportunistic feeders that eat most types of invertebrates they encounter in the leaf litter of the forest floor.

REPRODUCTION This is the only mole salamander (family Ambystomatidae) that breeds in streams. Males arrive at breeding sites in pools or flowing sections of limestone streams before females. After the females arrive, small groups of both sexes court and mate beneath large rocks over a period of several months, then leave to return to their terrestrial burrows in late winter to early spring. Females lay up to 260 eggs in a single layer on the underside of rocks, logs, or other submerged or partially submerged cover objects from late December to early April; several females may use the same rock. Some populations of streamside salamanders breed in small woodland pools, ditches, and flooded fields; and females attach eggs to logs, rocks, or grass stems.

PREDATORS AND DEFENSE The primary predators, especially of the larvae, are crayfish and green sunfish, although streamside salamanders do not usually occupy portions of streams where fish occur. Other predators include garter snakes and northern watersnakes. A type of flatworm has been known to prey on smaller individuals. Adults use biting, immobility, head hiding, and body posturing to avoid predation. Tail waving attracts predators to the part of the body where noxious compounds are secreted. Larvae hide under leaf litter and rocks in the stream bottom. Streamside salamanders are often active out in the open during daylight hours.

CONSERVATION The Tennessee Wildlife Resources Agency lists this species as Deemed in Need of Management and as an s2 species—a Natural Heritage ranking indicating that the species is imperiled because of its rarity. Conversion of forests and fields into housing developments threatens middle Tennessee populations. Stream pollution and habitat loss and alteration from deforestation threaten local populations throughout the range.

TAXONOMY The streamside salamander was first described in 1989. No subspecies are recognized.

The Y-shaped marking behind the eyes of this seepage salamander connects to the stripe along the middle of the back.

How do you identify a seepage salamander?

Seepage Salamander

Desmognathus aeneus

DESCRIPTION These tiny, slender salamanders have a round tail that makes up less than half the total length. A pale yellow, tan, or reddish brown, slightly wavy to straight-edged stripe usually bordered by dark brown occurs on the back and tail, although some individuals have a series of chevrons rather than a distinct stripe. A thin, dark, Y-shaped patch behind the eyes connects to a dark line or series of small herringbone-like lines that may continue into the stripe along the middle of the back. A light stripe runs from the eye to the jaw. The belly is uniform or mottled with brown and white.

Adult female with her eggs

LARVAE AND JUVENILES Hatchlings have a proportionally larger head and shorter tail than adults. Juveniles are patterned as adults, but their coloration is brighter. There is no aquatic larval stage.

BODY SHAPE
Tiny and slender

NUMBER OF COSTAL GROOVES
13 or 14

BODY COLOR
Brown

PATTERN
Usually a single light stripe with dark borders down the back

BODY SIZE
Adults to 2.5 inches in total length

This female seepage salamander represents one of the smallest species of salamander in North America north of Mexico.

SIMILAR SPECIES The top of the head is smooth in the seepage salamander but rugose in the pygmy salamander. The stripe on the back of Ocoee salamanders forms a zigzag pattern, and this species lacks a Y-shaped mark on the head. Southern red-backed salamanders lack the eye-to-jaw stripe and the Y-shaped mark.

DISTRIBUTION Seepage salamanders are found in the southern Appalachian Mountains from southeastern Tennessee and southwestern North Carolina through northern Georgia into Alabama. Disjunct populations occur in the Fall Line hills of central Alabama; the Piedmont of Georgia; and in Oconee County, South Carolina.

HABITAT These small salamanders are strongly associated with seeps and moist areas around small streams. They live between the moist leaf litter and the ground but may also be found under logs and mats of mosses. Although the highest population densities occur in seepage areas in cove forests, individuals occasionally move out into hardwood forests.

BEHAVIOR AND ACTIVITY Adults and juveniles remain hidden under leaf litter most of their lives. They do not aestivate and do not migrate.

FOOD AND FEEDING These diminutive salamanders forage under leaf litter, feeding on small invertebrates such as isopods, amphipods, earthworms, beetles, mites, and spiders. The observation of an adult male eating a juvenile indicates that cannibalism occurs.

seepage salamander
Desmognathus aeneus

REPRODUCTION Mating is terrestrial and occurs in February and March and from late July to October. After an elaborate courtship, the female picks up a spermatophore the male has deposited on the substrate. Females lay 3–19 eggs in moss or under decomposing logs in or near seeps, and remain with them until they hatch 6–9 weeks later. Eggs have been found from February to late May and hatch in late May to August.

PREDATORS AND DEFENSE Natural predators are unknown, but ringneck snakes, spring salamanders and other predatory salamanders, and leaf litter–foraging birds are likely to eat these little salamanders. Seepage salamanders of all sizes become immobile when they are uncovered, which probably makes them inconspicuous to predators.

CONSERVATION Populations are vulnerable to loss of mountain headwater streams and their associated seepages, but the seepage salamander is not protected in any state in the Southeast. Clearing of hardwood forests around headwater streams alters the seepage environment by eliminating leaf litter and moisture.

TAXONOMY The seepage salamander was first described in 1947. No subspecies are recognized.

Adult seepage salamanders are characterized by a broad brown stripe along the back and tail bordered on each side by a dark stripe.

A keeled tail flattened from side to side and orangish stripe behind the eye are characters of the southern dusky salamander.

How do you identify a southern dusky salamander?

BODY SHAPE
Robust

NUMBER OF COSTAL GROOVES
14

BODY COLOR
Dark brown to black

PATTERN
One or two rows of white spots on each side

BODY SIZE
Adults to 6 inches in total length

Southern Dusky Salamander

Desmognathus auriculatus

DESCRIPTION Southern dusky salamanders have a keeled tail that is flattened side to side and becomes more bladelike toward the tip, and a light eye-to-jaw stripe that is often orange-red. The back is dark brown to black with one or two rows of small white to reddish spots on each side between the front and hind limbs. Fine white specks may dot the back. Some individuals in peninsular Florida have a reddish back suggesting a light stripe. The belly is dark brown to black and sprinkled with small, distinctive white spots. Individuals from the western parts of the range tend to be lighter and sometimes have dark spots.

LARVAE AND JUVENILES Larvae are black to dusky brown with pale spots on the back, and their gills are dark and exceptionally bushy. Larvae in

Larvae are black to dusky brown with pale spots on the back.

southern dusky salamander
Desmognathus auriculatus

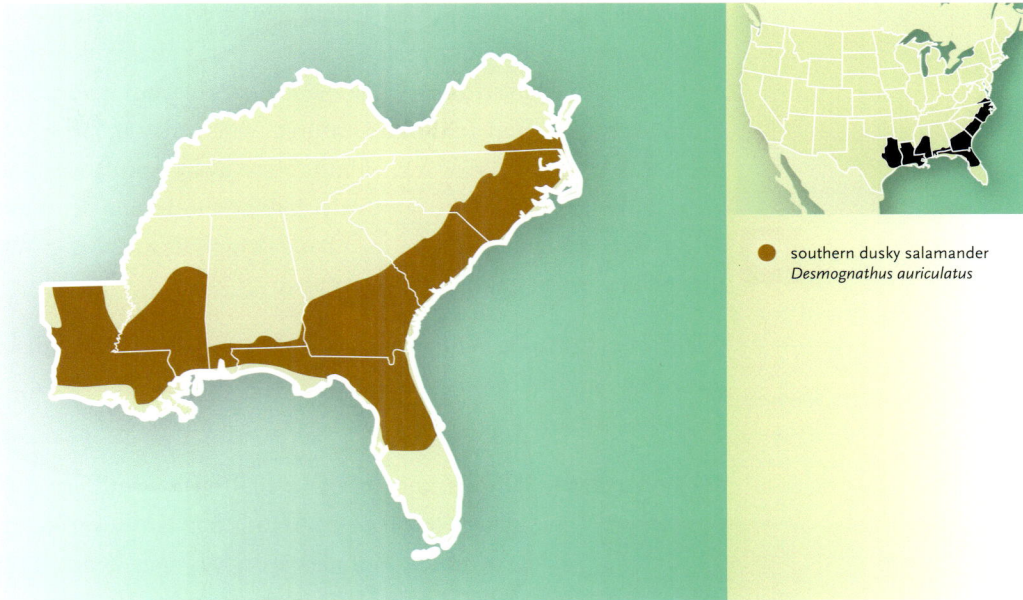

the Florida panhandle may have tiny white spots on the sides of the body and tail. Juveniles have six or seven pairs of light spots on the back.

SIMILAR SPECIES Spotted dusky salamanders have a cream to tan belly with irregular dark brown mottling; otherwise they are similar in color and pattern. Northern dusky salamanders often have a light, often reddish, irregular stripe on the back that extends onto the tail.

DISTRIBUTION Southern dusky salamanders are found in the Coastal Plain physiographic province from southeastern Virginia south to central Florida and west to Texas.

HABITAT These very aquatic salamanders are found in muddy wetlands in southern floodplain habitats and in swamps, seeps, sloughs, small, backwater streams, and cypress ponds. They live in deep, soupy, mucky organic substrates and can also be found under logs and leaf litter in seepage areas and mucky, acidic areas. They retreat beneath the peat layers when water levels are low.

A recently metamorphosed juvenile (top) and a young adult (bottom)

BEHAVIOR AND ACTIVITY The seasonal activity cycle is unknown, but these salamanders are likely to be active in their microhabitats as long as the surface water is not frozen. They tend to move around very little except during droughts and cold temperatures, when they burrow deeper into the muck and peat. They are seldom found away from seeps and mucky areas but will enter more terrestrial habitats to forage on rainy nights. Adults may defend territories as other salamanders in this group do, but that behavior has not been documented.

FOOD AND FEEDING Adults and juveniles eat beetles and their larvae, fly larvae, spiders, and worms. The larvae undoubtedly eat small crustaceans and other aquatic invertebrates.

REPRODUCTION Little is known about mating in southern dusky salamanders. They have been found in late summer in shallow water or muck. Females lay 9–26 eggs beneath objects on the muck surface, in sphagnum moss, and in cypress logs less than 6 feet from water in September to December. The larvae hatch in October to January, and metamorphosis takes place in January and February.

This southern dusky salamander found shelter from predators inside a dead snail shell.

PREDATORS AND DEFENSE Banded watersnakes and redfin pickerels are known to eat adults. Feral pigs rooting for food in drying wetlands probably eat them as well. Adults are cannibalistic and will eat eggs and juveniles. When exposed, these salamanders will try to escape quickly into muck, peat, and burrows.

CONSERVATION Many dusky salamander populations have declined dramatically since the 1970s. Loss of headwater and floodplain wetlands is the primary threat, although pollution and siltation damage water quality. Feral pigs constitute an underappreciated threat to populations that needs further evaluation. The reason why declines have occurred in parts of the range but not others also needs to be determined. Regulations covering collection for personal use—such as for fish bait—are the only protection this species receives in any state.

TAXONOMY Recent molecular studies indicate that the southern dusky salamander may actually comprise as many as five species. Future work may add several new species in the Southeast. First described to science in 1838, this species was long considered to be a subspecies of the northern dusky salamander. No subspecies are recognized.

The orange cheek and the line between the eye and back of the jaw combine to form the large, colorful blotch characteristic of many imitator salamanders.

Imitator Salamander

Desmognathus imitator

DESCRIPTION The imitator salamander is related to the mountain dusky group but is unique in having yellow, orange, or red cheeks in addition to the light (but sometimes dark) line between the eye and the angle of the jaw. The brown to black body and tail usually have a bold, strongly wavy-edged, light-colored stripe. Some old individuals lack both the colored cheek patches and the stripe and are a uniform grayish black. The belly of adults is uniformly gray. The tail, as in mountain duskies, is round at the base and lacks a keel.

LARVAE AND JUVENILES The juveniles are similar to adults in color and pattern. The larvae have not been described.

SIMILAR SPECIES The Ocoee salamander is most likely to be confused with the imitator. Those occurring within the range of the imitator salamander do not have colored cheeks, but they might be mistaken for older imitators that lack the colored cheek patches. If the stripe on the back is present, the imitator generally has a wavy-edged pattern and the Ocoee typically has a relatively straight-edged pattern. Imitator salamanders are not as abundant at high elevations as Ocoee salamanders are.

How do you identify an imitator salamander?

BODY SHAPE
Small with round tail

NUMBER OF COSTAL GROOVES
14

BODY COLOR
Gray, brown, black

PATTERN
Colorful cheeks; wavy-edged stripe on back

BODY SIZE
Adults to 4 inches in total length

imitator salamander
Desmognathus imitator

Look for these imitator salamanders at the edges of streams, in seepages, and on wet rock faces.

DISTRIBUTION Imitator salamanders are found in a small area in the Great Smoky, Plott Balsam, and Balsam mountains in eastern Tennessee and southwestern North Carolina.

HABITAT Imitator salamanders can be found on and under wet rocks at the edge of streams; in seepage areas; on wet rock faces; and in the leaf litter alongside headwater streams in spruce, fir, and hardwood forests at elevations above 2,200 feet.

BEHAVIOR AND ACTIVITY Imitator salamanders have been found only in warm-season months, but their habitats in the high mountains are very difficult for biologists to reach in winter. During the warm season the salamanders are nocturnal and seek cover during daytime. Adults disperse into nearby forested habitat during wet periods.

FOOD AND FEEDING Imitator salamanders feed at night from rocky perches, and possibly in the adjacent leaf litter as well. The diet has not been investigated but almost certainly includes a wide array of small invertebrates such as insects, worms, and spiders.

REPRODUCTION Mating probably occurs in spring and fall but has not been observed in nature. The male deposits a spermatophore on the moist substrate that the female picks up with her cloaca following an elaborate courtship. Females lay 13–30 eggs, attaching them to the underside of rocks and placing them in moss in moist areas alongside headwater streams. Nothing is known about the larval stage, but it may be similar to that of the Ocoee salamander and may last less than 1 year.

PREDATORS AND DEFENSE No predators have been documented, but ringneck snakes, garter snakes, watersnakes, and spring salamanders are probably among them. When discovered, the salamander either flees or becomes immobile. The red cheek color may mimic that of the Jordan's red-cheeked salamander, a species that visual predators consider distasteful.

CONSERVATION This high-elevation salamander may be subject to habitat and thermal changes associated with predicted climate change. Most populations are within Great Smoky Mountains National Park and are thus afforded some protection against habitat disturbances. Further studies must determine whether acid precipitation alters their habitat and food resources. The imitator salamander is not protected in North Carolina or Tennessee.

TAXONOMY The imitator salamander was first described in 1927 and was named "imitator" because it resembles another salamander in the region, the Jordan's red-cheeked salamander. No subspecies are recognized.

Little is known about the life history and ecology of this salamander, but advances in snow vehicle technology might allow scientists to study this species more completely.

Most Black Mountain salamanders are brown with irregular, sometimes faint, dark markings on the back. Their distribution is limited primarily to the Cumberland Mountains and Cumberland Plateau.

How do you identify a Black Mountain salamander?

BODY SHAPE
Slightly chubby; broad snout

NUMBER OF COSTAL GROOVES
14

BODY COLOR
Brown

PATTERN
Irregular black to dark brown spots or streaks

BODY SIZE
Adults to 6.75 inches in total length

Black Mountain Salamander

Desmognathus welteri

DESCRIPTION Black Mountain salamanders are brown with irregular black to dark brown spots or streaks on the back. The tail is flattened side to side, and its posterior half has a strong keel and is often lighter in color than the body. There is no sharp delineation between the color of the back and the sides. The belly is white with light brown to gray mottling. A light eye-to-jaw stripe is present, and the snout is broad and flattened. The tips of the toes are dark brown to black.

LARVAE AND JUVENILES Larvae and hatchlings have five to eight pairs of light spots on the back between the front and hind limbs that fuse and become obscure with age. The belly is white with light mottling that intensifies with age.

SIMILAR SPECIES Black Mountain salamanders are difficult to distinguish from seal salamanders and dusky salamanders in the same area. Seal salamanders do not have dark toe tips and usually have a yellowish to whitish area underneath the tail, and the color of their back is usually distinguishable from the color on the sides. The belly of seal salamanders lacks mottling, and that of the dusky salamander has a reticulated or salt-and-pepper pattern.

DISTRIBUTION Black Mountain salamanders in the Southeast are limited in distribution to the Cumberland Mountains and Cumberland Plateau of southwestern Virginia, southeastern Kentucky, and north-central Tennessee. They also occur in southwestern West Virginia.

HABITAT Adults and juveniles live in small to medium-sized streams with moderately steep gradients at elevations above 1,000 feet. They prefer streams with abundant rock cover that are shaded by forest canopy. Both age classes occupy wet stream margins and may wander into the adjacent forest during wet periods, although juveniles move farther than adults.

BEHAVIOR AND ACTIVITY These salamanders may be active in all but the coldest winter months; even then they may move around in water, but likely do not feed. They are not known to maintain territories but will move several yards up- or downstream to find suitable habitat. They emerge to forage along stream margins and in the adjacent forest on wet nights in warm months.

Larvae and hatchlings (below left) have five to eight pairs of light spots on the back between the front and hind limbs that fuse and become obscure with age.

Black Mountain salamanders (below right) have broad, flat snouts and a light eye-to-jaw stripe.

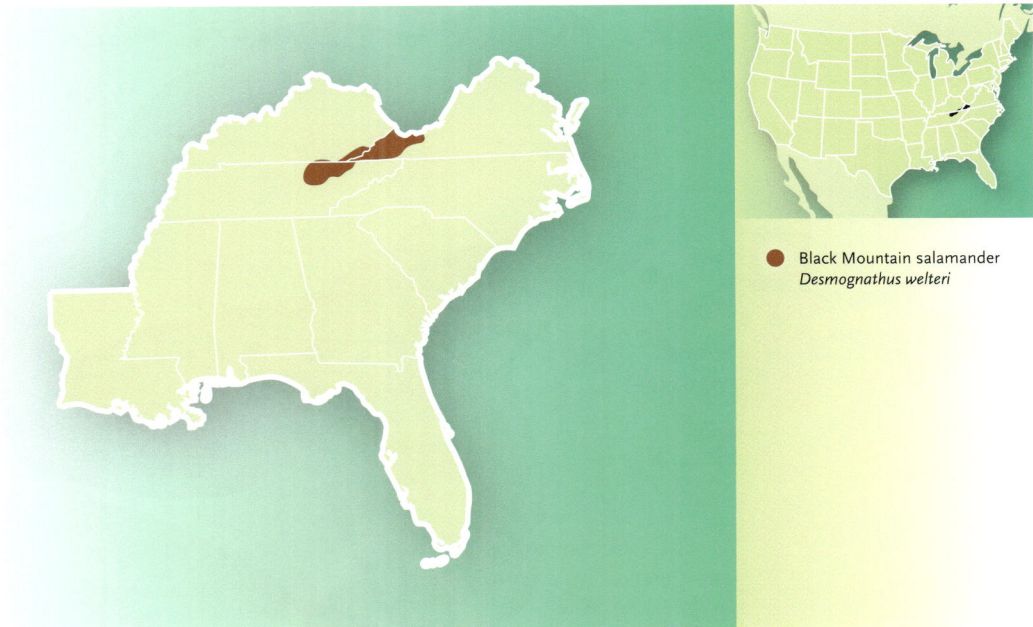

● Black Mountain salamander
Desmognathus welteri

FOOD AND FEEDING Adults and juveniles eat adult and larval flies, beetles, and larval and winged wasps and bees, suggesting that they forage for insects out of water while perched on rocks or in the adjacent forest. Captive salamanders will eat earthworms.

REPRODUCTION Little is known about courtship and mating behaviors, but they are likely to resemble the mating behaviors of other salamanders in this group. Females with spermatophores have been found in May. Females lay 18–23 eggs in grapelike clusters in small cavities they construct in packed leaves, under rocks in stream banks, and in wet crevices in small cascades. Mothers remain until their eggs hatch in late summer. Larvae metamorphose into juveniles the following summer.

PREDATORS AND DEFENSE No predators have been reported, but spring salamanders, garter snakes, and watersnakes are probably among them. Antipredator behaviors probably include biting, fleeing to rock shelters, and body twisting and flipping.

CONSERVATION The Black Mountain salamander is not protected by any state. It occurs in coal country, and so is subject to habitat loss and pollution from mining activities. Stream siltation and pollution from mine tailings are major threats. Mountaintop removal in which soil and rock are dumped downslope into streams destroys entire populations. Collection for fish bait by the "spring lizard" industry has been a serious threat but may be declining.

TAXONOMY The Black Mountain salamander was first described to science in 1950.

An adult Black Mountain salamander with typical color and pattern

How do you identify a black-bellied salamander?

Black-bellied Salamanders

Black-bellied Salamander   *Desmognathus quadramaculatus*

Dwarf Black-bellied Salamander   *Desmognathus folkertsi*

DESCRIPTION The common name of these large, stocky salamanders comes from the black belly that is present in adults and most juveniles. Adults are black to dark brown with irregular yellowish green to rusty blotches. In some populations the body may be covered with golden to brassy flecks. Old individuals may be nearly entirely black both above and below. The sides have a double row of conspicuous light spots. The snout and feet are usually light brown, and the toe tips are black. The tail is sharply keeled, less than half the body length, and flattened side to side. Dwarf black-bellied salamanders are almost identical but are about 30 percent smaller (to 6 inches in total length), have a pattern of alternating dark blotches on the back, and have a black belly when the body length is 2 inches or less. Both species have a light eye-to-jaw stripe.

BODY SHAPE
Stocky; tail less than half as long as body

NUMBER OF COSTAL GROOVES
14

BODY COLOR
Olive, reddish brown, dark brown, or black; black belly

PATTERN
Brown with irregular, rust-colored blotches

BODY SIZE
Adults to 8 inches in total length

Adults and most juveniles have black bellies.

This dwarf black-bellied salamander from Union County, Georgia, has a pattern of dark blotches on its back.

Black-bellied salamander larvae have small whitish gills but absorb them during metamorphosis 2–4 years after hatching.

LARVAE AND JUVENILES Larvae and juveniles of both species have a double row of light spots on the sides between the front and rear legs. Older larvae and juvenile black-bellied salamanders have a conspicuous reddish line down the top of the tail fin; the body of juveniles is often uniformly olive. Dwarf black-bellied salamanders lack the reddish line and are rarely greenish. Young juveniles have a white belly that becomes mottled with gray and darkens to completely black prior to sexual maturity.

SIMILAR SPECIES Older larval black-bellied salamanders have white gills and are larger than other *Desmognathus* larvae. Although hatchling black-bellied and dwarf black-bellied salamander larvae may be about the same size as the larvae of other species, the distinct rows of light side spots are distinctive. Other members of the genus usually do not have a black belly, although some old adults are black all over with a mottled belly and are difficult to distinguish. Shovel-nosed salamanders are mottled with much less black pigment, have a dusky belly that is not completely black, and the slope of the snout starts well behind the eyes.

DISTRIBUTION Black-bellied salamanders occur from southern West Virginia through western Virginia and along the Appalachian and Blue Ridge mountains of North Carolina and Tennessee south to Georgia. Dwarf black-bellied salamanders are known from most of Georgia's Blue Ridge and adjacent areas in North and South Carolina.

HABITAT These semiaquatic salamanders prefer high-gradient headwater and second-order mountain streams with abundant rock cover. They occa-

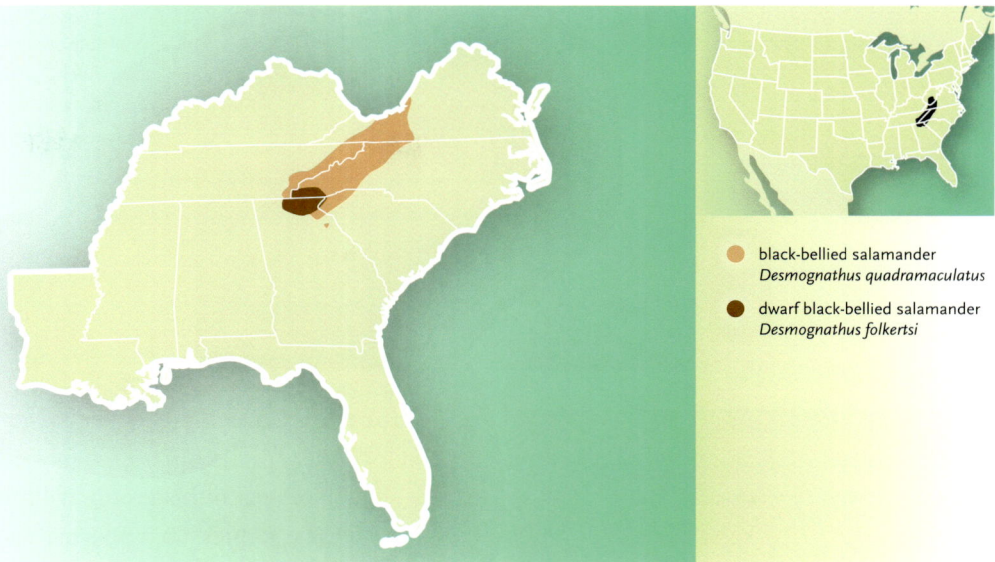

sionally leave the water to wander in the forest. Larvae live in high-current riffles, particularly those with gravel-sand interstices. Juveniles are most abundant in flat, rock-strewn edge habitats with numerous small hiding places. Adults, particularly large black-bellied salamanders, often use burrows in stream banks.

BEHAVIOR AND ACTIVITY Black-bellied salamanders are active year round if temperatures are warm enough to allow surface activity. Both species are adept at wedging themselves between rocks and enlarging holes in dirt. They defend such burrows and crevices and use them as bases from which to make occasional forays.

FOOD AND FEEDING Adults and juveniles feed on a wide variety of invertebrates in water, but during wet periods they may leave their streams to forage on and in the leaf litter up to 30 feet away. Larvae eat aquatic prey such as mayflies, stoneflies, craneflies, and crayfish. Juveniles and adults feed mostly on terrestrial prey, including bees and wasps, butterflies and moths, beetles, centipedes, and spiders. Larger salamanders can eat other salamanders.

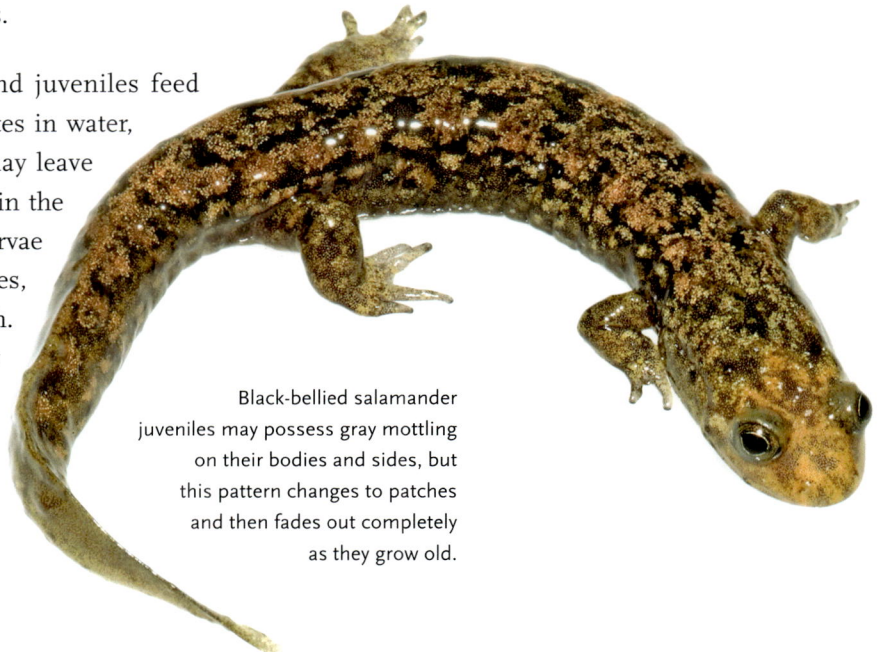

Black-bellied salamander juveniles may possess gray mottling on their bodies and sides, but this pattern changes to patches and then fades out completely as they grow old.

Black-bellied salamanders commonly prey on other salamanders, affecting community structure and the presence of other species.

Female black-bellied salamanders attach their eggs to the bottoms of rocks in shallow streams.

REPRODUCTION Courtship behavior has not been described for either species but is probably similar to that of other salamanders in this group, with the male enticing the female to pick up a spermatophore in her cloaca. Females lay as many as 80 eggs per clutch from April to July, attaching them to the underside of rocks in riffles or fast-moving streams as a single layer or in clumps. A female usually lays no more than one clutch every other year. The eggs hatch in summer and fall. The larvae of dwarf black-bellied salamanders take 2 years to reach metamorphosis; those of black-bellied salamanders take as long as 4 years.

PREDATORS AND DEFENSE Predators of black-bellied salamanders include shrews and garter snakes, but little is known about predator-prey relationships. Northern watersnakes, spring salamanders, aquatic invertebrates, and perhaps crayfish may be predators of larvae and small juveniles. Adults may on rare occasions bite a human finger but do not break the skin.

CONSERVATION The primary threats to black-bellied salamanders are alteration of natural streams resulting from acid precipitation and other atmospheric chemicals, and clear-cutting and lack of forest maintenance of watersheds above streams to prevent siltation and alteration of thermal qualities. Collection of these salamanders for fish bait (as "spring lizards") may once have caused population declines, but the practice has largely stopped and was probably never a widespread problem.

TAXONOMY The black-bellied salamander was described in 1840. The dwarf black-bellied salamander was described as a new species in 2002. No subspecies are recognized for either.

Did you know?

The primary prey of some salamanders is other salamanders. Black-bellied salamanders emerge from streams on rainy nights to forage on smaller species of salamanders along stream margins.

Body coloration is extremely variable among Ocoee salamanders from the same location and can change even more as individuals get older and larger.

How do you identify an Ocoee salamander?

Ocoee Salamander

Desmognathus ocoee

DESCRIPTION Ocoee salamanders have a light line between the eye and the back of the jaw. The base of the tail is slightly flattened (wider than tall), and the remainder is triangular. The pattern on the back is highly variable and usually includes a series of interconnected, yellow to orange, alternating, irregular spots with dark, wavy edging that extends well onto the tail. Some individuals have a straight-edged stripe. The body and tail are brown, gray, yellowish, or reddish and may be nearly black in old individuals. The belly is mottled with brown, gray, red, or yellow; old males may be nearly black. Some individuals have reddish cheeks or legs.

LARVAE AND JUVENILES Ocoee salamander larvae (to 0.7 inches in total length) and hatchlings (to 1.1 inches in total length) have seven pairs of conspicuous alternating light spots on a reddish brown back. Juveniles are brownish to reddish with alternating light spots edged in brown to black.

BODY SHAPE
Slender and slightly flattened

NUMBER OF COSTAL GROOVES
14

BODY COLOR
Brown to gray

PATTERN
Alternating light spots edged with black

BODY SIZE
Adults to 4 inches in total length

Ocoee salamander larvae and hatchlings have reddish brown backs with light spots.

SIMILAR SPECIES Spotted dusky salamanders have a keeled tail that is taller than wide at the base; the body is tan to brown with a wavy or smooth-edged yellowish to reddish brown stripe.

DISTRIBUTION Ocoee salamanders are found from southeastern Tennessee and southwestern North Carolina south of the Pigeon River into northern South Carolina and Georgia.

These small salamanders hide from predators in a wide variety of small places.

HABITAT Ocoee salamanders live in high-elevation seeps, first-order streams, wet rock faces, and mixed hardwood forests adjacent to streams. They avoid deep water and dry land, instead preferring spots where microhabitats interface. In hardwood forests they can often be found in wet mosses and under logs in moist leaf litter. Populations in low elevations (740 feet or less) are generally restricted to seepage areas and small streams; those in high elevations (to 4,600 feet) occupy the moist forest floor far from water.

BEHAVIOR AND ACTIVITY Ocoee salamanders are active on land in all but the coldest months. They remain hidden during the day beneath cover or in rock face crevices and emerge at night to feed and search for mates. Adults actively defend their small territories against other adults. Adults and juveniles congregate deep in seeps and underground retreats in winter.

Ocoee salamander
Desmognathus ocoee

Some Ocoee salamanders
have red legs.

Females lay 12–28 eggs in
July or August and stay with
the eggs until they hatch in
August or September.

FOOD AND FEEDING Ocoee salamanders will eat virtually any small animal they can capture and swallow, including their own eggs and hatchlings. Prey of adult and juveniles includes flies, ants, beetles, wasps, spiders, mites, grasshoppers, and moths. Adults will climb and search for food in low, usually herbaceous, vegetation on wet nights.

REPRODUCTION Courtship and mating occur on land in spring and in late summer and fall. In July and early August, females lay 12–28 eggs in small clusters under rocks in seeps, in leaf litter, under moss, and in crevices in wet rock faces where shallow running water is present. Mothers remain with their eggs until they hatch in August to late September after a developmental period of 2–2.5 months. The larval period is 8–10 months; metamorphosis occurs in May and June, when the larvae are about an inch long.

PREDATORS AND DEFENSE Known predators of Ocoee salamanders include ringneck snakes, garter snakes, watersnakes, black-bellied salamanders, seal salamanders, spring salamanders, small mammals, and birds that forage in leaf litter. When threatened they may autotomize their tail, flip their body, become immobile, or flee. Some of the more colorful individuals in this group may receive some protection by mimicking toxic newt efts.

Populations of this highly variable species occur on isolated mountain ranges. Modern taxonomic techniques may reveal that some of these are separate species.

Ocoee salamanders often climb vegetation at night during moist conditions to feed on arboreal invertebrates.

CONSERVATION These small salamanders are considered to be among the most abundant salamanders in the southern Appalachians and are not protected except where they occur on protected lands. Their fragile habitats are threatened, however, by timber removal and climate change that may alter hydrological patterns in the mountains. Continued monitoring of selected areas along elevational gradients may reveal population changes in response to regional climate changes.

TAXONOMY The Ocoee salamander is part of the mountain dusky salamander complex and was first described in 1949. Numerous subgroups within this species occupy different mountain ranges, and each may warrant species recognition in the future.

The stout body, large hind legs, and light jaw stripe are typical of adult Apalachicola dusky salamanders. The tail is round at the base and tapers significantly at the tip.

How do you identify an Apalachicola dusky salamander?

BODY SHAPE
Small; tail rounded and filamentous

NUMBER OF COSTAL GROOVES
14

BODY COLOR
Brown to nearly black

PATTERN
Series of reddish round spots

BODY SIZE
Adults to 4 inches in total length

Apalachicola Dusky Salamander

Desmognathus apalachicolae

DESCRIPTION Apalachicola duskies are robust salamanders with rear legs that are larger than the front legs. A light stripe runs between the eye and jaw. The long tail is round at the base and tapers into a filament. Five to seven pairs of somewhat round cream, tan, red, or yellowish blotches run down the middle of the back and onto the tail on a black to brown background. The body spots may fuse to form a light stripe bordered on each side by black or dark brown. The top of the tail is the same color as the blotches. The head and sides of the body and tail are brown. Adult males, especially old ones, may be entirely black or brown. The belly is uniformly white, although some individuals have smudges of dark pigment. Females are more vividly patterned than males.

LARVAE AND JUVENILES The larvae have five to seven pairs of light blotches with tiny, white, sticklike external gills. Juveniles are colored and patterned as adults with yellow, brown, or red blotches but are brighter.

SIMILAR SPECIES The southern dusky is nearly uniformly black to dark brown with small white or reddish flecks concentrated on the sides and on the black belly. Spotted dusky salamanders have a brown back with black markings in an irregular stripe edged in black and a cream to tan belly with dark brown mottling.

DISTRIBUTION Apalachicola dusky salamanders are found in southeastern Georgia from Upson County south to the center of the Florida panhandle.

HABITAT These salamanders inhabit seeps and small headwater streams in the Chattahoochee and Ochlockonee river watersheds. They hide under leaves, logs, and rocks during the day and move and forage at night. They often sit in debris piles in very shallow water with only their head exposed in the air. The larvae remain in very shallow flowing water. Apalachicola duskies are rarely found away from water.

Apalachicola dusky salamanders occur primarily in the moist environments of steep ravines associated with seepages and headwater streams from southeastern Georgia to the center of the Florida panhandle.

BEHAVIOR AND ACTIVITY These salamanders are known to be primarily nocturnal, but their seasonal activity patterns are unknown. They may be active year round, although they retreat underground in the hot summer months and move about on the surface only during rain. Adults are seldom found together, suggesting that they maintain territories.

FOOD AND FEEDING Larvae and adults are entirely carnivorous and probably feed on small insects and other invertebrates.

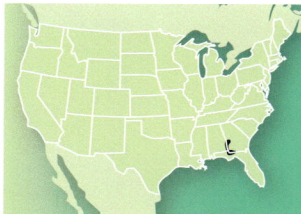

● Apalachicola dusky salamander
Desmognathus apalachicolae

Apalachicola dusky salamanders hide under leaves, logs, and rocks during the day and move about at night to forage.

REPRODUCTION The timing of courtship and mating is not known. Females lay loose clusters of eggs under rocks, small logs, or other cover objects on stream edges or in seeps in May and June. Clutch size has not been reported. Females remain with their eggs until they hatch in May through November, presumably to deter potential predators.

Juveniles are lighter and have more distinct markings than adults.

PREDATORS AND DEFENSE No predators have been documented, but they likely include several species of snakes, other salamanders such as red salamanders, snapping turtles, large frogs, shrews, raccoons, and other mammals. Apalachicola dusky salamanders rely on concealment and cryptic coloration to avoid predators. They can also leap by using their strong, enlarged rear legs.

CONSERVATION The topography in the range of this small salamander affords some measure of habitat protection because the deep ravines are not suitable for houses or other buildings, although damming a stream to make a pond can destroy the habitat. The species has received no formal protection because it remains locally abundant.

TAXONOMY The Apalachicola dusky salamander was not found and described until 1989. No subspecies have been described.

Spotted dusky salamanders occur from the Appalachian highlands to the Gulf of Mexico.

How do you identify a common dusky salamander?

BODY SHAPE
Robust with rear legs larger than front legs

NUMBER OF COSTAL GROOVES
14

BODY COLOR
Brown to gray

PATTERN
Highly variable, often with a broad reddish stripe on the back

BODY SIZE
Adults to 5.5 inches in total length

Common Dusky Salamanders

Northern Dusky Salamander	*Desmognathus fuscus*
Spotted Dusky Salamander	*Desmognathus conanti*
Santeetlah Dusky Salamander	*Desmognathus santeetlah*
Virginia Dusky Salamander	*Desmognathus planiceps*

DESCRIPTION Northern dusky salamanders and the three closely related species described here are medium-sized (to 5.5 inches in total length), stout-bodied salamanders with rear legs larger than the front legs. The base of the tail is taller than wide, and the remainder is triangular in cross section and keeled along the top edge. The eye-to-jaw stripe is prominent in all but old individuals. The base color is brown to gray. The back pattern, if present, is highly variable and darkens with age. Reported patterns include six or seven pairs of black, irregularly shaped spots and a continuous dark yellowish to reddish brown stripe that may be straight-edged or wavy along each edge. The reddish stripe on the back usually extends onto the base of the tail. The belly is yellowish brown to cream with some black to gray mottling.

Spotted dusky salamanders have bolder markings on the back, and the spots may be red to golden. Santeetlah dusky salamanders have a poorly defined stripe on a brownish to greenish background; the underside of the tail may be yellowish. Virginia dusky salamanders differ from northern duskies in tooth size (*D. planiceps* has broader crowns and slightly larger teeth) but otherwise are nearly identical. The differences are primarily genetic.

LARVAE AND JUVENILES The larvae are small (less than an inch long) with five to eight pairs or sets of alternating light spots on a brownish background; external gills are present. Juveniles retain the spots until maturity, at which time they fuse into the broad stripe.

SIMILAR SPECIES Mountain dusky salamanders have a rounded tail. The dark pattern on the back of seal salamanders is reticulate (netlike) or a series of irregular black spots. Black-bellied salamanders have a black belly. Shovel-nosed salamanders have a paddle-like tail and two rows of light blotches on the back. Larval and juvenile desmogs are the most difficult to identify. Some species can be eliminated on the basis of locality. Consult the descriptions of each species in this book and study photographs in several sources to further narrow the possibilities.

DISTRIBUTION Northern dusky salamanders range from Nova Scotia south through Kentucky and Virginia into the Piedmont, the Blue Ridge and Appalachian mountains, and the Cumberland Plateau. They overlap and may hybridize with spotted dusky salamanders in South Carolina, northern Georgia, and Tennessee. Spotted duskies occur from the Appalachian highlands west to the Mississippi River delta and south to the Gulf of Mexico on the Florida panhandle, Alabama, Mississippi, and Louisiana; isolated populations occur in the Georgia Coastal Plain, north-central Louisiana, and northeastern Arkansas. Santeetlah dusky salamanders are found in the Great Balsam Mountains of North Carolina and in the Great Smoky and Unicoi mountains of North Carolina and Tennessee. Virginia dusky salamanders are found in the Blue Ridge Escarpment east into the southern Piedmont of Virginia and possibly North Carolina.

Santeetlah dusky salamanders occur in three mountain ranges in North Carolina, including the Great Smoky Mountains.

Northern dusky salamander larvae have white gills that are absorbed during metamorphosis.

Northern dusky salamanders range north to Nova Scotia.

Common dusky salamanders (above, from left to right), northern dusky salamander, spotted dusky salamander, Santeetlah dusky salamander, Virginia dusky salamander

HABITAT Dusky salamanders prefer cool, well-oxygenated streams and their associated seepages. In the lowlands, this salamander appears to be restricted to seepage areas along small streams and the edges of swamps and slow-moving streams. Northern and Santeetlah salamanders will disperse into the surrounding forest in rainy periods and may be found under logs and rocks.

BEHAVIOR AND ACTIVITY Dusky salamanders are active during all but the coldest winter months, when they move into rocky crevices deep below the surface. From spring through fall they hide during the daytime under leaf litter, rocks, and logs in seeps and along stream margins, and forage and move at night. Communication is visual and by means of pheromones, but these salamanders do not mark territories. They remain under cover when relative humidity is low or it is not raining; the drier it is, the larger the rock they choose to hide under. High humidity and night rain bring out insects and other prey, and the salamanders come out to feed on them. Some individuals disperse up to 50–60 feet into the surrounding forest when the substrate is wet.

FOOD AND FEEDING Larvae feed on copepods, fly and other insect larvae, collembolans, mites, and small fingernail clams. Juveniles and adults eat any animal they can catch and swallow. The larger the salamander, the larger the prey item it can consume. Aquatic and terrestrial insects and other invertebrates such as larval and adult flies, caddisflies, true bugs, beetles, ants, spiders, caterpillars, mites, and earthworms are common prey types.

REPRODUCTION Timing of courtship, egg laying, and other aspects of the life history of this wide-ranging salamander vary with elevation in the Southeast and with latitude in northern areas. Mating occurs in spring and fall. Some females appear to lay their eggs in the fall following mating; some do not lay eggs until the following May, June, or July. Each female

126 • *Common Dusky Salamanders*

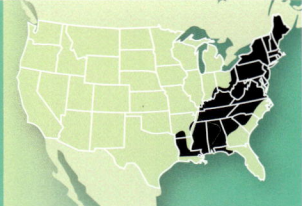

- northern dusky salamander
  Desmognathus fuscus
- spotted dusky salamander
  Desmognathus conanti
- both northern dusky and
  spotted dusky salamanders

- Virginia dusky salamander
  Desmognathus planiceps
- Santeetlah dusky salamander
  Desmognathus santeetlah

Common dusky salamanders lay one to three dozen eggs in grapelike clusters in small depressions. The female remains with the eggs until they hatch.

deposits 11–36 eggs in grapelike clusters under rocks in water, in cavities under and inside logs, in leaf mats, or in clumps of moss within 6 feet of the stream edge. Santeetlah dusky salamanders lay about 3 more eggs per female than northern dusky salamanders. Females in some areas lay eggs every other year. Larvae hatch in about 2 months and remain in shallow water with a gravel substrate to avoid the cannibalistic adults. The length of the larval period depends on time of hatching, water temperature, and food availability. Larvae that hatch in spring may metamorphose by the following fall, but those that hatch in late summer or fall overwinter as larvae and metamorphose more than a year later. Size at metamorphosis is 1–2 inches.

PREDATORS AND DEFENSE Known predators include northern watersnakes, common garter snakes, shrews, raccoons, skunks, bears, and black-bellied, red, and spring salamanders. Cannibalism of larvae and adults occurs, but its frequency is unknown. Dusky salamanders will bite potential predators but will first attempt to flee. The tail will break off if a predator bites it instead of the body.

CONSERVATION Loss of headwater and seepage habitats as a result of agriculture, urbanization, and road building has caused loss and decline of populations, especially those outside protected areas such as state or national parks and forests. Clear-cutting of hardwood trees affects water availability and thermal characteristics in seeps and small streams. Collection of "spring lizards" for fish bait can greatly reduce the size of some populations. Populations of these long-lived (about a decade) salamanders that lay relatively few eggs cannot tolerate the removal of many adults. Mountaintop mining operations are always a threat because they increase siltation and destroy habitat. Very little is known of any aspect of the life history and population ecology of the species in this dusky salamander group. Understanding the variation in timing of reproductive and other critical events could provide insight into how to manage populations as climate changes alter their habitats. Details of the ecology and life history of the Virginia dusky salamander are almost completely unknown.

TAXONOMY The northern dusky salamanders were described first in 1820. The spotted dusky was initially described as a subspecies of the northern dusky in 1958 but was elevated to a full species in 2002. The Santeetlah dusky was described in 1981. The Virginia dusky was first described in 1955 but was not recognized as a full species until 2008. No subspecies are recognized for any species in this group.

The black-and-brown body pattern, large pop eyes, and tail with a keel on the posterior portion are typical of seal salamanders.

Seal Salamander

Desmognathus monticola

DESCRIPTION Seal salamanders are stout-bodied, pop-eyed animals with a flattened tail that has a sharply defined keel on the last two-thirds of the upper surface. The brown to dark gray body has a bold pattern of black or brown markings that are wormlike, netlike, or in circles surrounding areas of background color. The belly is plain grayish cream without markings. A single row of small, irregularly shaped white spots may be present on each side between the front and hind legs. Older adults tend to lose the pattern on the back completely and turn purplish brown while their belly becomes gray or brown. Some populations may have dark brown or black toe tips.

LARVAE AND JUVENILES Larvae have small white gills and four or five pairs of light spots on a light brown body. Hatchlings and juveniles are brown with four or five pairs of chestnut or orange-brown spots along the back. This pattern changes quickly into the adult form within a year.

How do you identify a seal salamander?

BODY SHAPE
Robust; knife-edged tail tip

NUMBER OF COSTAL GROOVES
14

BODY COLOR
Brown to gray

PATTERN
Irregular bold, dark reticulation or spots

BODY SIZE
Adults to 5 inches in total length

Larval seal salamanders (left) have small white gills and 4 or 5 pairs of light spots on a light brown body. Juveniles (right) are brown with 4 or 5 pairs of chestnut or orange-brown spots along the back.

SIMILAR SPECIES Northern dusky salamanders and other desmogs in that group have a salt-and-pepper belly and usually a reddish brown edge to the stripe on the top of the base of the tail. Black-bellied salamanders have a black belly as adults, but juveniles have yellow flecking; both age classes have two rows of tiny white dots along each side. The back of shovel-nosed salamanders is mottled to resemble the pebble substrate in streams, and the sloping of the head starts behind the eyes; the head of seal salamanders starts sloping at the eyes. Black Mountain salamanders have a white belly with darker mottling.

DISTRIBUTION In the Southeast, seal salamanders are found from western Virginia and eastern Kentucky southward through the Appalachian and Blue Ridge mountains and northern Georgia to south-central Alabama.

Seal salamanders exhibit a wide range of variation and may be dark (top) or light (bottom) in background color.

HABITAT Many cool freshwater streams with well-aerated water in hardwood forests in the Appalachian and Blue Ridge mountains and foothills harbor seal salamanders. Mountain streams with a rocky substrate and gravel bars are preferred over silted, slow-moving streams. Salamanders avoid lower reaches of streams that support fish, and juveniles remain in gravel areas to avoid the cannibalistic adults.

BEHAVIOR AND ACTIVITY Adults emerge from the water at night and sit on the rocks they hide under in the daytime to wait for prey. Juveniles minimize predation by avoiding the large rocks used by adults. Seal salamanders are active year round in water, even when there is ice on the surface. During periods of heavy rains, some adults leave their stream and move a short distance into the adjacent forest, presumably to forage. During rain they occasionally climb 3–6 feet on tree trunks to forage. Males actively defend their shelter sites from other males.

FOOD AND FEEDING Adults sit atop their rock shelters at night and wait for invertebrates such as bugs, beetles, flies, ants, stoneflies, moths, beetles, crickets, mayflies, millipedes, and earthworms to come along. Juveniles actively move around to forage, but the adults usually stay in one place.

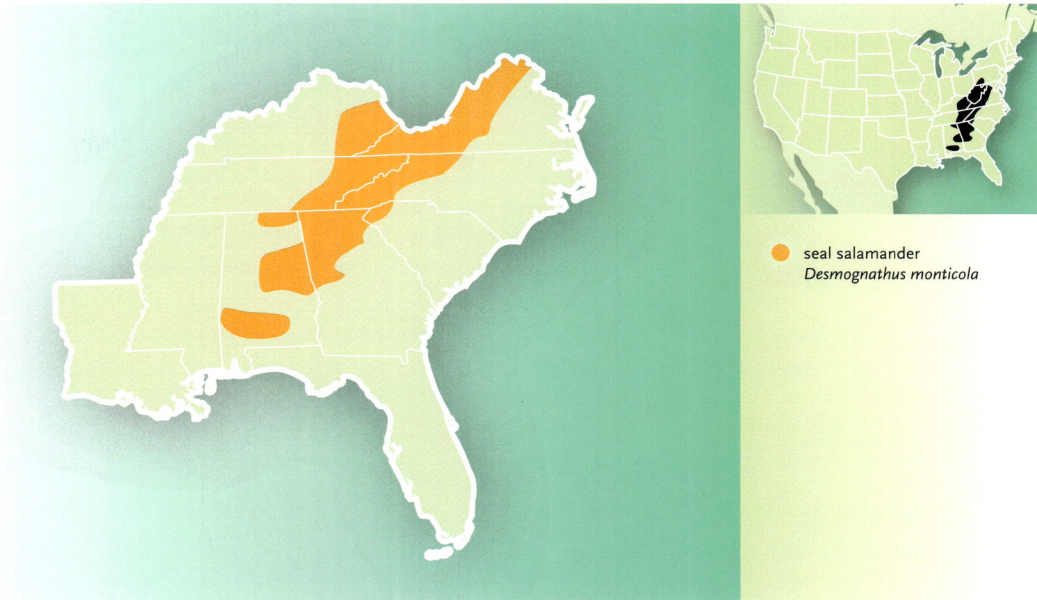

seal salamander
Desmognathus monticola

REPRODUCTION Mating occurs in spring and fall. Females produce 13–40 eggs, apparently annually, during mid-June to August and lay them as loose clumps or in a single layer attached to the underside of rocks in streams, in moss in water, in wet crevices, and in seepage banks. Eggs hatch in late summer to early fall, and the larvae overwinter in the stream. The external gills are absorbed 8–10 months later during metamorphosis.

PREDATORS AND DEFENSE Spring salamanders, several species of fish such as brook trout, watersnakes, and ringneck snakes eat seal salamanders. Their main defenses are their slippery skin and rapid escape by diving into the stream to hide under a rock.

CONSERVATION Fishermen collect seal and other stream salamanders to use for bait ("spring lizards"), and overcollection in some areas may have contributed to the decline of this species. The collection of salamanders for fish bait is prohibited in Jefferson National Forest in Virginia. Acid precipitation has apparently caused their decline in some streams. Clear-cutting around mountain streams alters the stream environment and contributes to population declines.

TAXONOMY The seal salamander was described initially in 1916. No subspecies are recognized.

Seal salamanders are slippery and escape quickly by diving into streams and hiding under rocks. Older individuals like this one may be uniformly dark with little pattern.

Some Blue Ridge dusky salamanders have a straight-edged, reddish stripe down the back.

How do you identify a mountain dusky salamander?

BODY SHAPE
Slender with a rounded tail

NUMBER OF COSTAL GROOVES
14

BODY COLOR
Gray, brown, or yellow

PATTERN
Light stripe bordered by black

BODY SIZE
Adults to 4 inches in total length

Mountain Dusky Salamanders

Cumberland Dusky Salamander	*Desmognathus abditus*
Carolina Mountain Dusky Salamander	*Desmognathus carolinensis*
Allegheny Mountain Dusky Salamander	*Desmognathus ochrophaeus*
Blue Ridge Dusky Salamander	*Desmognathus orestes*

DESCRIPTION These four closely related species look so much alike that knowing where the salamander came from is critical for accurate identification. The tail is round at the base, and the flattened posterior portion lacks a keel. A light line connects the eye and angle of the jaw. The pattern on the back ranges from a straight-edged, light-colored (light brown, gray, reddish, yellowish, or red) stripe to a series of interconnected, light-colored blotches with dark, wavy edging. The stripe usually extends onto the base of the tail. Narrow black chevrons or irregular black spots may be present inside the light stripe. The body and tail are brown to nearly black, and the belly is light in young salamanders and light gray to black in adults. Old males may be completely black. The straight and wavy-edged patterns occur in all four species; the straight-edged form predominates in the Allegheny mountain dusky. The eight pairs of reddish spots on the back of Cumberland dusky salamanders are usually in evidence even in older, dark adults.

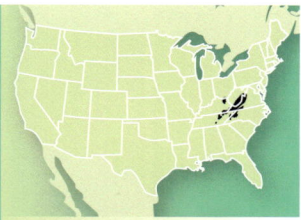

- Cumberland dusky salamander
  Desmognathus abditus
- Carolina mountain dusky
  salamander *Desmognathus carolinensis*
- Allegheny Mountain dusky
  salamander *Desmognathus ochrophaeus*
- Blue Ridge dusky salamander
  Desmognathus orestes

The light line from the eye to the back of the jaw of this Carolina mountain dusky salamander is the signature trait of the dusky salamander group.

Although many of the dusky salamanders are similar in appearance, Cumberland dusky salamanders often have reddish spots on the back.

This Allegheny Mountain dusky salamander has black chevrons or irregularly shaped black spots inside the light stripe down its back.

Mountain dusky salamanders occur in cool, high-elevation seepage areas in mixed hardwood forests near streams. Wet mossy areas are a common habitat.

The paired light spots of hatchling dusky salamanders change to stripes or blotches as they become juveniles.

LARVAE AND JUVENILES Cumberland dusky salamander larvae and juveniles are unknown. Larval Allegheny and Carolina mountain dusky salamanders have a rounded snout, external gills, and five or six pairs of alternating black spots on the back. Juveniles usually have reddish or yellowish stripes or blotches on the back. Hatchling Blue Ridge dusky salamanders have not been described, but juveniles have brownish, reddish, or yellowish stripes.

SIMILAR SPECIES The tail of northern dusky salamanders, seal salamanders, and their close relatives is taller than wide at the base and has a keeled edge. Members of this group also tend to be more robust than the mountain dusky group. The tail of the closely related Ocoee salamander is wider than high at the base and appears somewhat flattened, and a series of alternating light spots bordered by black scalloping forms an irregular wavy or straight line down the back. Two-lined salamanders are yellow with two dark lines along the back, and they lack the light eye-to-jaw stripe, as do all of the woodland salamanders in the genus *Plethodon*.

DISTRIBUTION Cumberland duskies are known only from three counties in the Cumberland Plateau in central Tennessee. Carolina mountain dusky salamanders are found in the Blue Ridge, Black, Bald, and Unaka mountains in western North Carolina and eastern Tennessee. Allegheny mountain dusky salamanders occur in the northeastern Allegheny Mountains and in the Southeast from western Virginia south through eastern

Tennessee. Blue Ridge dusky salamanders are limited to the Blue Ridge Mountains in southwestern Virginia, northwestern North Carolina, and northeastern Tennessee.

HABITAT Mountain dusky salamanders occupy cool, high-elevation seepage areas, first-order streams, wet rock faces, and mixed hardwood forests adjacent to streams. They avoid deep water and prefer the interface between flowing water and dry ground such as that found in wet mosses and under logs in moist leaf litter in forests. Allegheny mountain dusky salamanders are often highly terrestrial and can be found great distances from water. Juvenile mountain dusky salamanders are found most often in damp leaves and along stream margins.

BEHAVIOR AND ACTIVITY Mountain dusky salamanders are active on or near the surface in all but the coldest months—that is, generally from March through October. They remain active in seepage areas and springs even in winter. Their movement from streams and seeps into the surrounding forests coincides with rainfall; in dry conditions they limit their activity to wet habitats. Low-elevation populations remain close to seepages and wet rock faces, but those in high elevations wander into the surrounding forest.

FOOD AND FEEDING The timing of metamorphosis coincides with the abundance of small invertebrates in shallow water. Juveniles and adults are generalist predators that consume a wide variety of prey, including ants, beetles, grasshoppers, caddisflies, spiders, mites, moths, and worms. Adults occasionally eat eggs, hatchlings, and juveniles of their own species.

Juvenile mountain dusky salamanders like this one are found most often among damp leaves and along stream margins.

REPRODUCTION Mating occurs in water or in moist spots on land from September through June in all but the coldest months. Females lay 5–37 eggs in grapelike clusters under rocks and logs; in moss, mud, and seepage banks; and in crevices on wet cliffs with flowing water. Timing of hatching varies with elevation, latitude, and food availability. Hatchlings have been found in fall and spring. The larval period likewise can be short (1–3 weeks) or long (8–10 months). Metamorphosis, consisting primarily of loss of gills, occurs about 7–10 months later, typically around May and June, again depending on location.

PREDATORS AND DEFENSE Known predators include ringneck snakes, garter snakes, watersnakes, spring salamanders, small mammals such as shrews, and turkeys and other birds that forage in leaf litter. These salamanders may bite, break off the tail, become immobile, or flee when threatened. Some of the more colorful individuals in this group may receive some degree of protection from mimicking toxic newt efts.

Mountain dusky salamanders typically remain active on or near the surface from March through October.

CONSERVATION A large portion of the ranges of these four species in the Southeast occurs on public lands such as state and national parks, forests, and wildlife refuges. However, clearing land for timber threatens the environmental integrity of the headwater streams and seeps these salamanders require. Subsequent siltation and removal of the protective canopy can result in higher temperatures. Mountaintop removal eliminates entire populations. Climate change that increases temperatures in high elevations will also threaten seepage areas and other microhabitats. None of these salamanders are protected legally except by state regulations that limit the number that can be collected. The small range of the Cumberland dusky salamander makes that species vulnerable to habitat loss or alteration, and its conservation status needs immediate attention. Local climate change is expected to cause mountain species such as these to migrate upslope and reduce their ranges, and local populations may be extirpated.

TAXONOMY Until 1996 these four species were considered one species, the Allegheny mountain dusky salamander, described in 1859. The Carolina mountain dusky salamander was described in 1916 as a subspecies, the Blue Ridge dusky salamander in 1996, and the Cumberland dusky salamander in 2003. No subspecies are currently recognized. The species in this complex can be distinguished from each other only by geographic location and genetic analysis.

The colorful southern two-lined salamander is found in all of the states in the Southeast.

How do you identify a two-lined salamander?

Two-lined Salamanders

Northern Two-lined Salamander   *Eurycea bislineata*

Southern Two-lined Salamander   *Eurycea cirrigera*

DESCRIPTION Adults of both species are usually less than 4 inches in total length but can be almost 5 inches long. They are yellow, sometimes brownish yellow, with black or dark brown lines that run down the sides and onto the tail. Tiny black spots may be visible on the back. Individuals in some populations are uniformly brown. The belly is translucent and plain yellow. During the breeding season males commonly have a pair of cirri extending downward from the outer edge of the upper lip below the nostril. Northern and southern two-lined salamanders can be differentiated by the number of costal grooves (the northern species typically has 15 or 16, and the southern typically has 13 or 14) and by location; their ranges overlap in the Southeast only in Virginia.

BODY SHAPE
Small and slender

NUMBER OF COSTAL GROOVES
13–16

BODY COLOR
Yellow or light brown

PATTERN
Dark stripe on either side of the back

BODY SIZE
Metamorphs 1.6 inches; adults to 4.7 inches in total length

LARVAE AND JUVENILES The aquatic larvae are tiny at hatching (less than 0.5 inch) and have obvious red gills. Larvae are dull gray or brownish with light spots along the sides and

The tiny aquatic larvae of two-lined salamanders have light spots along the sides instead of lines.

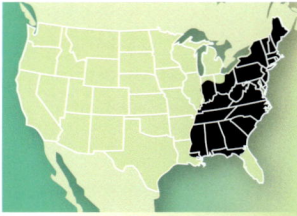

northern two-lined salamander
Eurycea bislineata

southern two-lined salamander
Eurycea cirrigera

Northern two-lined salamanders are often brownish yellow with a dark line down each side and tiny black spots on the back.

back and a clear belly. The end of the head is blunt. A clear tail fin is visible from above the hind legs to the end of the tail. Juveniles resemble adults.

SIMILAR SPECIES The Blue Ridge two-lined salamander is often more orange, and the dark stripes become broken about halfway down the tail, but they remain complete for most of the tail length in the northern and southern two-lined salamanders; geographic distribution is an effective means of differentiating most specimens because of the limited areas of overlap. Three-lined salamanders have a dark line down the middle of the back, and the dark side stripes of long-tailed salamanders are more broken and irregular than those of two-lined salamanders. Adult three-lined and long-tailed salamanders are larger than either of the two-lined species.

Most dusky salamanders that share habitats with two-lined salamanders have a brown or black body without distinct dark stripes and have a light mark behind the eye.

DISTRIBUTION Two-lined salamanders are widespread in the Southeast. The ranges of the southern and northern species merge in Virginia in a narrow zone running from southern Rockingham County in the Shenandoah Valley northeastward to southern Prince William County in the Coastal Plain.

HABITAT Two-lined salamanders are associated with small streams or seepage areas bounded by forest habitat. Adults are most frequently seen in late winter and early spring in wooded stream habitats, often far from the stream. Both species occur in urban and suburban streams if rocks and other hiding places are abundant.

BEHAVIOR AND ACTIVITY Some adults may be present under streamside rocks and logs at any time of year, but they are especially prevalent in March and April during the breeding season. They are most active in early evening. Juveniles and adults may live in drier forested areas far from the water, although they usually remain under moist ground cover or in burrows. Adults have been reported to be territorial. Some two-lined salamander populations spend most of the year in wooded habitats and migrate to stream pools for breeding.

FOOD AND FEEDING These small salamanders eat a wide variety of terrestrial invertebrates including earthworms, sowbugs, millipedes, beetles, snails, flies and midges, wood roaches, spiders, and ticks; and aquatic prey such as caddisfly larvae, stonefly nymphs, and fish eggs.

REPRODUCTION Two-lined salamander reproductive activity has been observed in most months of the year and varies according to geographic location, habitat type, and seasonal rainfall and temperature. Males in some populations develop cirri prior to breeding. Whether courtship and mating are primarily terrestrial or aquatic is unknown for most populations. A female will lay as few as 15 to more than 100 eggs, which she attaches beneath rocks or logs in

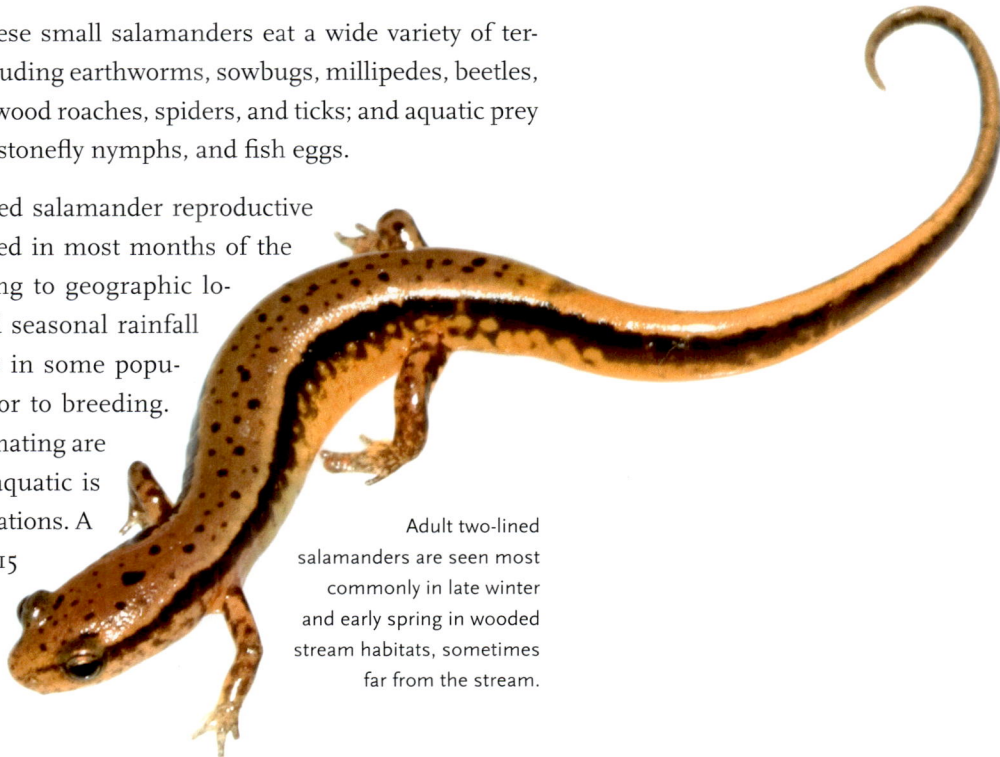

Did you know?

Senescence is not known to occur among salamanders. Some species are known to live more than a decade in the wild and more than a quarter of a century in captivity.

Adult two-lined salamanders are seen most commonly in late winter and early spring in wooded stream habitats, sometimes far from the stream.

A female two-lined salamander may lay up to 100 eggs, attaching them beneath rocks or logs in slow-moving water. She remains with her eggs until they hatch.

slow-moving water. She remains with the eggs until they hatch, which may be about a month to more than 2 months. The larvae remain in the water for more than a year until they metamorphose.

PREDATORS AND DEFENSE Numerous predators have been documented, including ringneck snakes and garter snakes, screech owls, black-bellied salamanders, and trout. Defensive maneuvers include remaining immobile, running, and tail waving. The tail will break off if a predator grabs it, allowing the salamander to escape. Secretions from skin glands make two-lined salamanders distasteful to some predators, such as certain species of shrews.

CONSERVATION Two-lined salamanders are geographically widespread and are usually present or even abundant in woodland stream habitats in urbanized areas. Local populations are susceptible to activities that result in excessive stream pollution, heavy siltation, or degradation of the terrestrial landscape.

TAXONOMY The northern two-lined salamander was described in 1818, and the southern two-lined salamander in 1830. The southern two-lined salamander was considered a subspecies of the northern two-lined salamander until 1987, as was the Blue Ridge two-lined salamander (*E. wilderae*). A variant of the southern two-lined salamander found in north-central Alabama has been proposed as a new species called the brown-backed salamander (*Eurycea aquatica*), but this designation has been controversial and requires additional research.

Blue Ridge two-lined salamanders have black spots on the back and well-defined dark lines that run down the sides and onto the tail.

How do you identify a Blue Ridge two-lined salamander?

Blue Ridge Two-lined Salamander

Eurycea wilderae

DESCRIPTION Blue Ridge two-lined salamanders are bright yellow, yellowish orange, or orange. Broad, well-defined dark lines run down the sides and onto the tail, breaking up midway down the tail into blotches. The back has black spotting. The belly is plain yellow. The total length is commonly 3.5 to more than 4 inches and may reach 4.75 inches. Males often have cirri and enlarged head muscles during the breeding season.

LARVAE AND JUVENILES The aquatic larvae have gills and a noticeable clear keel on the tail; they are about 0.4 inch long at hatching and slightly more than an inch long at metamorphosis. The larvae can be dull brown, gray, or greenish yellow with light spots on the back and sides. They metamorphose after 1 year in warmer streams but may take up to 2 years in colder ones. The juvenile stage lasts about a year. Juveniles resemble adults.

BODY SHAPE
Small and slender

NUMBER OF COSTAL GROOVES
14–16

BODY COLOR
Bright yellow to orange

PATTERN
Dark stripes on sides that break into blotches about halfway down the tail

BODY SIZE
Metamorphs 1.1 inches; adults to 4.75 inches in total length

Blue Ridge two-lined salamander larva

Blue Ridge two-lined salamanders often occur in hardwood forests and spruce forests, sometimes several hundred feet from streams.

SIMILAR SPECIES The Blue Ridge two-lined salamander can be distinguished from the southern two-lined salamander in areas where the two might overlap geographically by the latter's less well-defined side stripes, which usually remain unbroken on the tail. Other salamanders within the species' geographic range include the much larger three-lined salamander, which has a dark center line down the back; and the long-tailed salamander, whose side stripes are not as well defined. Other salamanders found in the habitats where Blue Ridge two-lined salamanders live do not have the bright yellow or orangish body color along with distinct side stripes. Dusky salamanders are mostly brown or black and have a light mark behind the eye.

DISTRIBUTION Blue Ridge two-lined salamanders range through the southern Appalachians in the Southern Blue Ridge Mountain physiographic province from White Top Mountain, Virginia, across North Carolina, and into about a dozen counties in northeastern Georgia, including areas within the Upper Piedmont province. The range is completely surrounded by that of the southern two-lined salamander.

HABITAT These salamanders occupy mountain stream systems in mixed hardwood forests and spruce forests to elevations above 6,000 feet. They often occur in the terrestrial habitat, sometimes several hundred feet from streams.

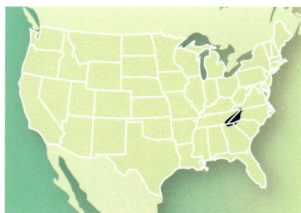

Blue Ridge two-lined
salamander
Eurycea wilderae

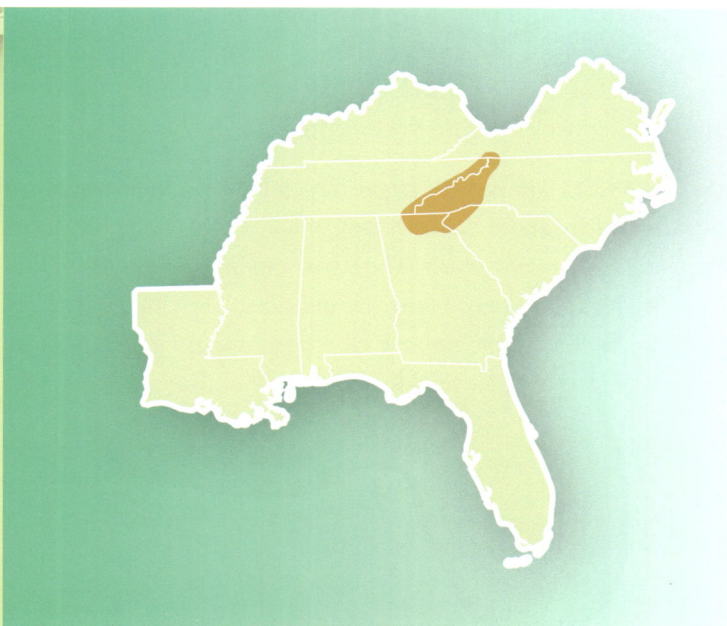

Blue Ridge two-lined salamanders occur in parts of Virginia, North Carolina, and Georgia.

BEHAVIOR AND ACTIVITY During nonbreeding seasons Blue Ridge two-lined salamanders sometimes migrate far into forested areas and remain under ground litter such as leaves and dead bark. They can be found beneath rocks along mountain streams at most times of the year but are less abundant during summer.

FOOD AND FEEDING Blue Ridge two-lined salamanders probably eat a wide variety of small terrestrial and aquatic invertebrate prey such as earthworms, sowbugs, flies, spiders, and the larvae of aquatic insects.

REPRODUCTION The timing of reproduction varies according to elevation, stream temperature, and other factors not yet fully determined. As early as October adults begin migrating from woodland habitats to streams where they congregate for courtship and mating. Some males develop prominent cirri during the breeding period; the head becomes enlarged in others. Each female lays a clutch of 8–56 eggs in late winter to early spring, gluing them to the underside of rocks in the water. Several females and egg clusters can sometimes be found under the same rock. The female remains with her eggs until they hatch in late spring to early summer, the timing depending on water temperature and time of egg laying.

PREDATORS AND DEFENSE Larger salamanders, including spring salamanders and black-bellied salamanders, and garter snakes are known predators. Red salamanders, shovel-nosed salamanders, and ringneck snakes probably eat them, also. Defensive strategies are variable and probably include remaining immobile when exposed; rapid retreat under cover objects; and waving the tail, which can break off.

CONSERVATION Blue Ridge two-lined salamanders have historically been among the most common stream salamanders within their geographic range and continue to thrive in most areas that have not suffered undue aquatic pollution or siltation. Thermal changes to environments at high elevations that could alter stream temperatures might affect populations of these salamanders.

TAXONOMY The Blue Ridge two-lined salamander was first described in 1920 as a subspecies of the northern two-lined salamander and was elevated to full species status in 1987.

Female Blue Ridge two-lined salamanders remain with their eggs until they hatch in late spring to early summer.

Three-lined salamanders are recognizable by the broad, distinct dark line running the length of the body and tail on each side and the dark line down the center of the back.

How do you identify a three-lined salamander?

BODY SHAPE
Long and slender; broad head

NUMBER OF COSTAL GROOVES
13 or 14

BODY COLOR
Dull yellow to tan

PATTERN
Dark stripe on each side and down center of body

BODY SIZE
Metamorphs 1.2 inches; adults to 7.9 inches in total length

Three-lined Salamander *Eurycea guttolineata*

DESCRIPTION The body is usually yellow but can be tan, brownish, or orangish. A broad, distinct dark brown or black line runs the length of the body and tail on each side. The line may become infused with lighter color on the tail but is still apparent. A dark line, which may be broken or well defined and continuous, extends down the center of the upper body from behind the head to the base of the tail. The belly is mottled with gray pigment on a cream to yellow background. Adults are often more than 6 inches in total length and can be nearly 8 inches long. The tail is long, sometimes making up almost two-thirds of the total body length.

LARVAE AND JUVENILES The gilled aquatic larvae are about 0.5 inch long at hatching and are grayish to cream-colored with stripes and a clear tail fin. The larvae metamorphose at a length of 0.9–1.3 inches after 4–6 months in warmer areas, or after more than a year in colder ones. Juveniles resemble adults in pattern and color.

SIMILAR SPECIES The three dark longitudinal body stripes on a yellowish body distinguish the three-lined salamander from other species within the geographic range. Some long-tailed salamanders have a central row of spots down the back and can resemble three-lined salamanders, but the two species do not overlap in most parts of their respective ranges.

three-lined salamander
Eurycea guttolineata

DISTRIBUTION The geographic range of the three-lined salamander includes the portion of Virginia east of the Blue Ridge Mountains and most of the Carolinas and Georgia to the Florida panhandle, most of Alabama and Mississippi to the upper southeastern corner of Louisiana, western Tennessee, and southwestern Kentucky. The species is noticeably absent from most of the Coastal Plain of North Carolina, the southeastern and northwestern corners of Georgia, and all of peninsular Florida.

Three-lined salamanders' bellies are mottled gray on a cream to yellow background.

HABITAT Three-lined salamanders are found in seepage areas, stream edges, and in river floodplains and swamps. They are sometimes found in the crevices of rock outcrops some distance from water. They are primarily terrestrial but do not venture far from the aquatic sites where they lay their eggs, and they remain in moist areas beneath ground litter of leaves, logs, and dead bark, or under rocks.

BEHAVIOR AND ACTIVITY These salamanders are active above ground at night, especially during rain, and remain beneath cover during the day. Individuals are most often seen during warmer months but can be found under ground litter throughout the year. Unlike many other terrestrial sala-

Three-lined salamanders are widespread geographically, being found in parts of every southeastern state, and are abundant in many areas.

manders that live in wooded areas and breed in aquatic habitats, three-lined salamanders usually stay in the vicinity of the water. They are not overtly territorial as many other stream salamanders are.

FOOD AND FEEDING The diet consists of a wide variety of small terrestrial invertebrates, including several types of insects such as ants, flies, beetles, grasshoppers, and others; spiders; snails; worms; and millipedes.

REPRODUCTION Surprisingly little is known about the courtship and mating patterns of three-lined salamanders, which are relatively abundant over a wide geographic range. Eggs are laid in the late fall or winter, mainly in December, but apparently not in the large identifiable clusters characteristic of other stream salamanders. Females are believed to lay about a dozen eggs underwater but do not stay with them until hatching.

PREDATORS AND DEFENSE Direct predation has not been observed, but ringneck snakes, garter snakes, and watersnakes would readily eat these salamanders, as would larger salamanders that live in the same habitat. If they are unable to escape by running or swimming away, three-lined salamanders sometimes assume a defensive behavior by hiding the head beneath the tail, which they curl and slowly wave.

CONSERVATION Three-lined salamanders are widespread geographically and abundant in many areas. Local populations may be susceptible to habitat alterations such as the clearing of bottomland hardwood forests, siltation, and pollution.

TAXONOMY The three-lined salamander was described in 1838 and was considered a subspecies of the long-tailed salamander (*Eurycea longicauda*) until 1998, when genetic research indicated that three-lined and long-tailed salamanders do not normally hybridize in regions of overlap.

Long-tailed salamanders typically have black dots on the back and sides, no central stripe, and a long tail.

How do you identify a long-tailed salamander?

Long-tailed Salamander

Eurycea longicauda

DESCRIPTION The basic body color is yellow, often with a tinge of orange or brown. The belly is yellowish without a pattern. Black dots are present on the back and sides; those on each side often form a distinct dark line, but those on the back do not. The dark herringbone pattern on the sides of the tail is characteristic. The long tail can be almost two-thirds of the total body length. Both sexes may have cirri, but they are more prominent in males.

LARVAE AND JUVENILES The aquatic, gilled larvae are 1.5–2 inches long and are gray or cream-colored with a clear tail fin. Duration of the larval period depends on water temperature and other environmental conditions; it can be as short as 2 months but is usually about 6 months. Larval periods of more than a year and possibly 2 years have been reported. Juveniles resemble adults, but their black pigment is not as pronounced.

BODY SHAPE
Small and slender; broad head

NUMBER OF COSTAL GROOVES
13 or 14

BODY COLOR
Orangish yellow or brownish yellow

PATTERN
Black spots on body that may be in rows but do not form distinct lines

BODY SIZE
Metamorphs 1 inch; adults to 7.75 inches in total length

Long-tailed salamander larvae are aquatic with noticeable gills.

SIMILAR SPECIES The long-tailed salamander is most likely to be confused with the three-lined salamander, which has a distinct stripe down the middle of the back; and the cave salamander, which has a reddish body and scattered black spots on its body and tail. The spots on the tail of the long-tailed salamander, in contrast, coalesce to form a series of Z-shaped transverse bars (herringbone pattern) down the length of the tail.

DISTRIBUTION The geographic range of the long-tailed salamander in the Southeast includes all of Kentucky and western Virginia; northern Georgia, Alabama, and Mississippi; and the eastern two-thirds of Tennessee. The range is generally outside that of the three-lined salamander, although some overlap occurs.

HABITAT Long-tailed salamanders are mostly terrestrial and are generally associated with seepage areas and the margins of small streams and seasonal wetlands, although individuals are occasionally found in wooded habitats far from water. They typically hide beneath logs, rocks, and other ground cover during the daylight hours and forage at night. The species is especially noted for its association with caves and habitats where shale is present.

BEHAVIOR AND ACTIVITY Long-tailed salamanders associated with surface environments or twilight zones of caves are nocturnal and are especially active on rainy nights during warm periods of the year. Adults readily swim.

long-tailed salamander
Eurycea longicauda

Congregations of numerous individuals are not uncommon in some areas, suggesting that the species is not territorial. Individuals living in forests migrate to aquatic sites during the breeding season, and populations have been observed migrating in and out of caves and mine shafts.

FOOD AND FEEDING Adults eat almost every kind of small invertebrate in their habitat: insects such as beetles, flies, and moths; spiders; millipedes; isopods; worms; and springtails. Larvae eat small aquatic invertebrates such as midge and mayfly larvae, ostracods, copepods, and snails.

REPRODUCTION Long-tailed salamanders mate from early fall through winter, with the actual months of reproductive activity determined by geographic location, altitude, and seasonal temperatures and rainfall patterns. Mating is presumably aquatic, and a female may lay from 60 to as many as 100 eggs from fall to spring. The eggs are usually laid singly or in clusters that become scattered, and the female does not remain with the clutch.

Long-tailed salamanders (above) are mostly terrestrial and are often associated with caves and habitats where shale is present.

Female long-tailed salamanders (left) lay more than 60 eggs, singly or in clusters that become scattered. The female does not remain with the eggs after they are laid.

PREDATORS AND DEFENSE Sculpins and sunfish are known aquatic predators of long-tailed salamanders, and terrestrial predators probably include snakes, larger salamanders, and shrews. When threatened, adults run to cover or wave the tail, which breaks off if a predator grabs it.

CONSERVATION The species is widespread and often abundant and does not appear to be seriously threatened on a broad scale. Local populations have probably been reduced or possibly extirpated in situations where strip mining or acid runoff from mine tailings has altered stream habitats.

TAXONOMY The long-tailed salamander was described in 1818. Three subspecies were recognized until 1998, when genetic studies revealed the three-lined salamander (*Eurycea guttolineata*) to be a distinct species. The dark-sided salamander (*E. longicauda melanopleura*) of the Ozark highlands and the eastern form (*E. l. longicauda*) are still considered a valid subspecies.

Junaluska salamanders have small, irregular dark brown spots on the back and sides that form wavy, disconnected stripes on a yellowish brown body.

How do you identify a Junaluska salamander?

BODY SHAPE
Small and slender

NUMBER OF COSTAL GROOVES
14

BODY COLOR
Yellowish brown

PATTERN
Dark brown mottling on back; wavy stripes on sides

BODY SIZE
Metamorphs about 1.6 inches in body length; adults to 4 inches in total length

Junaluska Salamander   *Eurycea junaluska*

DESCRIPTION The body is yellowish brown with small, irregular darker brown spots on the back and sides. The markings on the sides form wavy, disconnected stripes. The belly is light greenish yellow and unmarked. The tail constitutes about half the total length. During the breeding season, males may develop a small mental gland and have short cirri and enlarged head muscles.

LARVAE AND JUVENILES The gilled, aquatic larvae are olive to brown with light spots on the back and sides and a clear tail fin. They are about 0.5 inch long at hatching and about 3 inches long at metamorphosis, which occurs after 2–3 years. The larvae are essentially adult size at metamorphosis.

SIMILAR SPECIES The Blue Ridge two-lined salamander, which is found in the same habitat, is more orange than the Junaluska and has distinct stripes partway down the sides. Dusky salamanders have a distinctive light line from the eye to the jaw.

DISTRIBUTION The geographic range of the Junaluska salamander is restricted to Graham County in western North Carolina and adjacent counties in eastern Tennessee.

HABITAT Junaluska salamanders are normally found under rocks and logs along margins of streams bordered by hardwood forest habitat.

BEHAVIOR AND ACTIVITY Adults and juveniles leave their hiding places at night to hunt; adults are often found crossing roads on rainy nights.

FOOD AND FEEDING The diet is not known but presumably consists of a variety of small aquatic and terrestrial invertebrates.

REPRODUCTION Males may develop small cirri, mental glands, and enlarged head muscles during the breeding season, but courtship and mating have not been observed. In the spring, females lay clusters of 30–68 eggs (average 38) in streams more than a foot deep, attaching them to the underside of rocks. The female remains with the eggs until they hatch after about a month, usually in April, May, or June. Metamorphosis occurs in May to August.

PREDATORS AND DEFENSE No predators have been documented, but larger salamanders, snakes, shrews, and birds would presumably regard these little salamanders as prey.

CONSERVATION The species is susceptible to timbering and other activities that affect stream water quality. Because of their limited geographic distribution and relative rarity in most habitats where they occur, local populations are in danger of being eliminated if their streams and the forests surrounding them are not properly managed. The Junaluska salamander is listed as a Threatened species in North Carolina.

TAXONOMY The Junaluska salamander was first recognized as being distinctive from the two-lined salamander in 1937 and was described as a new species in 1976.

Junaluska salamander larvae are aquatic and have gills. The body is olive to brown with light spots on the back and sides.

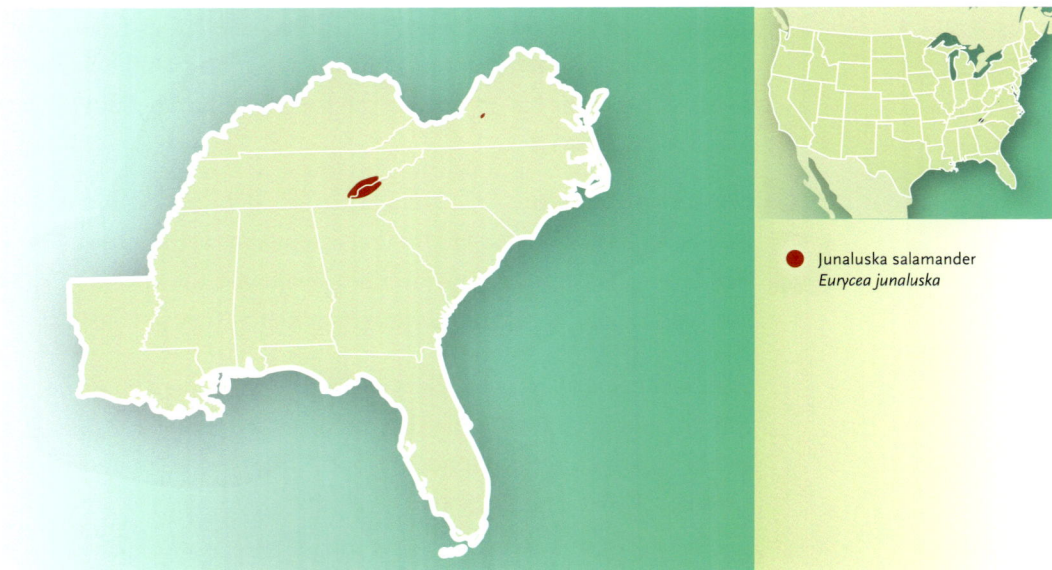

● Junaluska salamander
Eurycea junaluska

This Chamberlain's dwarf salamander from Aiken, South Carolina, is a female with eggs. The presence of only four toes on each back foot is a key character of the dwarf salamanders.

How do you identify a dwarf salamander?

BODY SHAPE
Tiny and slender

NUMBER OF COSTAL GROOVES
15–18

BODY COLOR
Yellowish to brown

PATTERN
Dark stripe on each side

BODY SIZE
Metamorphs 0.8 inches; adults to 3.5 inches in total length

Dwarf Salamanders

Dwarf Salamander   *Eurycea quadridigitata*

Chamberlain's Dwarf Salamander   *Eurycea chamberlaini*

DESCRIPTION These small, slender salamanders have a yellowish to brownish back bordered by dark stripes along the sides that continue onto the tail. The dwarf salamander has a gray or silver belly; Chamberlain's dwarf salamander has a yellow belly. The two species differ from most other southeastern salamanders in having only four toes on the back feet. The dwarf salamander has 17 or 18 costal grooves; Chamberlain's dwarf salamander has 15 or 16.

LARVAE AND JUVENILES Hatchlings are less than 0.3 inch in total length, with red gills and a clear tail fin that extends forward in front of the hind legs. Young larvae and young juveniles are grayish brown above. The young have spots and assume the pattern of adults as they get older.

SIMILAR SPECIES The two species of dwarf salamanders might be confused with the southern two-lined salamander, which reaches a larger body size and has five toes on each hind foot; or the four-toed salamander, which has a white belly with black spots.

DISTRIBUTION The geographic distribution is a bit confused because the dwarf salamanders were considered a single species for more than 150 years. One or the other, or both, occur in the Coastal Plain from North Carolina south through most of Florida and west to eastern Texas and southern Arkansas. The range of Chamberlain's dwarf salamander appears to be more restricted, being known to occur for certain in the Carolinas, Georgia, and Alabama, and possibly Florida. It has also been found in portions of the Piedmont in North and South Carolina. The two species are found together in some areas of the Carolinas and Georgia.

HABITAT Dwarf salamanders are common in and around many wetland habitats, including Carolina bays, seepage areas, ponds in pine flatwoods, and swamp margins. They are sometimes active above ground during the day when migrating.

Dwarf salamanders typically have a yellowish body with dark stripes along the sides and tail. Adults are usually less than 3.5 inches long.

Dwarf salamander larvae are tiny with a grayish brown body and red gills (left).

Chamberlain's dwarf salamander (right) has a yellow belly.

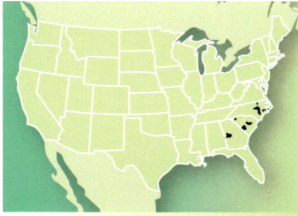

Chamberlain's dwarf
salamander
Eurycea chamberlaini

dwarf salamander
Eurycea quadridigitata

BEHAVIOR AND ACTIVITY Both species of dwarf salamanders leave their forested habitat or the margins of wetland areas for the wetland itself during the breeding season in the fall, and return to terrestrial habitats in the spring. They are most commonly found beneath leaves, logs, pine straw, and other ground litter.

FOOD AND FEEDING Small invertebrate prey items include numerous kinds of insects, spiders, mites, and millipedes.

REPRODUCTION Dwarf salamanders move to aquatic breeding sites from late summer into fall; Chamberlain's dwarf salamanders move later in the season, from November to February. The timing within a population of either species varies geographically and in response to seasonal weather conditions. The female lays 40–45 eggs, usually attaching them singly to the underside of vegetation in the water. Some females lay eggs in low, dry areas of wetlands that will fill with fall and winter rains. Dwarf sala-

Dwarf salamanders and Chamberlain's dwarf salamanders usually remain motionless when picked up, allowing an observer to count the number of costal grooves and distinguish between the species.

In both Chamberlain's dwarf salamander (top) and the dwarf salamander (bottom) the back is bordered by dark stripes along the sides that continue onto the tail.

Dwarf salamanders leave their forested habitat or the margins of wetland areas for the wetland itself during the breeding season in the fall and return to terrestrial habitats in the spring.

manders are not known to remain with the eggs, but Chamberlain's dwarf salamanders have been reported to do so. The larvae metamorphose in as few as 2 months or as many as 6.

PREDATORS AND DEFENSE Pig frogs and black swamp snakes are known predators of dwarf salamanders, and a variety of semiaquatic snakes, frogs, salamanders, and even predaceous insects and spiders probably prey on them as well.

CONSERVATION Both species can be abundant in some areas, but local populations can be extirpated if wetlands are lost. The dwarf salamander is listed as a Species of Special Concern in North Carolina.

TAXONOMY The dwarf salamander was originally described in 1842. Three subspecies were subsequently described but are no longer recognized. Chamberlain's dwarf salamander was noted in 1991 as a color morph occurring in many of the same locations as the dwarf salamander and was described as a separate species in 2003.

A slender, orange body and wide, flat head with large eyes are trademark traits of the cave salamander.

Cave Salamander

Eurycea lucifuga

DESCRIPTION These long, slender salamanders have a wide, flat head; large eyes; a long, prehensile tail; and a body strikingly colored in orange, reddish orange, or yellowish orange. The body and tail are covered with large, irregularly placed black spots. The belly is whitish with no dark markings.

LARVAE AND JUVENILES The aquatic, gilled larvae are about 0.5 inch in total length at hatching and are pale brown with lengthwise rows of light spots on the sides. Juveniles are yellowish but become reddish as they mature.

SIMILAR SPECIES The cave salamander is most likely to be confused with the long-tailed salamander, but the spotting on the sides of the tail of the latter coalesces into a herringbone pattern. Long-tailed salamanders are also more yellowish than orange and do not have the flattened head characteristic of cave salamanders.

Hatchling cave salamanders are pale brown with lengthwise rows of light spots on the sides.

How do you identify a cave salamander?

BODY SHAPE
Slender; wide, flat head; large eyes

NUMBER OF COSTAL GROOVES
14 or 15

BODY COLOR
Orange

PATTERN
Large black spots on body and tail

BODY SIZE
Metamorphs 2.5 inches; adults to 7.1 inches in total length

Juvenile cave salamanders (left) are yellowish but become reddish as they mature (right).

DISTRIBUTION Cave salamanders are found from just north of Roanoke, Virginia, southwestward along the border of western Virginia and eastern West Virginia south to northwestern Georgia, and westward to northern Alabama and northeastern Mississippi. They also occur in the eastern three-fourths of Tennessee, in most of Kentucky and into Indiana, and west to Oklahoma.

HABITAT The cave salamander is associated with karst (limestone) topography, and most populations are found in and around caves or rocky cliff habitats, commonly in the twilight zone. Large numbers have been found hundreds of yards inside caves, but individuals are also found in wooded habitats in the vicinity of rock outcrops, caves, and abandoned mine shafts.

BEHAVIOR AND ACTIVITY Cave salamanders are noted for climbing on and within rock faces and crevices, using their prehensile tail for balance. They are often found beneath rocks, logs, leaves, or other ground litter, and may be active on the surface during rains.

FOOD AND FEEDING Cave salamanders are known to prey on more than 100 different kinds of invertebrates, including flies and beetles, springtails, isopods, millipedes, mites, earthworms, snails, spiders, centipedes, crayfish, and snails; they also eat juvenile slimy salamanders.

Cave salamanders are associated with karst (limestone) topography, and most populations are found in and around caves or rocky cliffs.

PREDATORS AND DEFENSE Potential predators include other salamanders, snakes, shrews, and birds. Cave salamanders will hide the head beneath the body and wave the tail, which contains toxic compounds and will detach if grabbed by a predator.

REPRODUCTION Courtship and mating are presumed to occur during summer and early fall, with eggs being laid from September to February, the timing varying among localities. Single eggs are laid in rocky stream pools inside or outside caves. Each egg is attached to the bottom, and a female may lay up to 120 eggs (average more than 60). The larval period is at least a year and may be 15–18 months.

CONSERVATION Although cave salamanders are widespread and abundant in some areas, populations that are strongly associated with cave systems are vulnerable if the surrounding forest habitat or water quality of streams within or outside caves is degraded.

TAXONOMY The cave salamander was described as a species in 1822. No subspecies are recognized.

Cirri, which appear during the mating season, are visible on this male cave salamander.

The irregular black streaks and spots on this spring salamander are typical for the species. The yellow line from the eye to the tip of the snout marks the canthus rostralis.

How do you identify a spring salamander?

BODY SHAPE
Robust; truncated head; tail fins present

NUMBER OF COSTAL GROOVES
17–19

BODY COLOR
Salmon to pink

PATTERN
Small streaks, spots, or chevrons

BODY SIZE
Metamorphs 3–5 inches; adults to 9 inches in total length

Spring Salamander   *Gyrinophilus porphyriticus*

DESCRIPTION Spring salamanders are stout bodied with irregular black streaks, spots, or reticulation on a salmon to yellowish brown to reddish orange body, head, and tail. A light white to yellow line usually bordered by gray to black extends along the canthus rostralis from the eye to the tip of the broad snout. The belly is pinkish and unmarked. The tail is flattened from side to side. Individuals become darker with age.

LARVAE AND JUVENILES The larvae have a long, truncated snout; external gills; and a flattened tail with fins. Body color varies from pink to light yellowish brown to light gray with black flecking. The larvae become quite large (to 6 inches) before they metamorphose into adults in 4–5 years. They lack the light or dark spots found on the larvae of other salamanders in the region.

SIMILAR SPECIES No other salamander in the range of this species has a light line on a rounded snout. Dusky salamanders and related species have a prominent eye-to-jaw stripe. Mud salamanders are reddish brown with scattered distinct, round black dots. Red salamanders are reddish with numerous irregular black dots that run together with age. Tennessee cave salamanders have tiny eyes and a longer head and wider snout than spring salamander larvae.

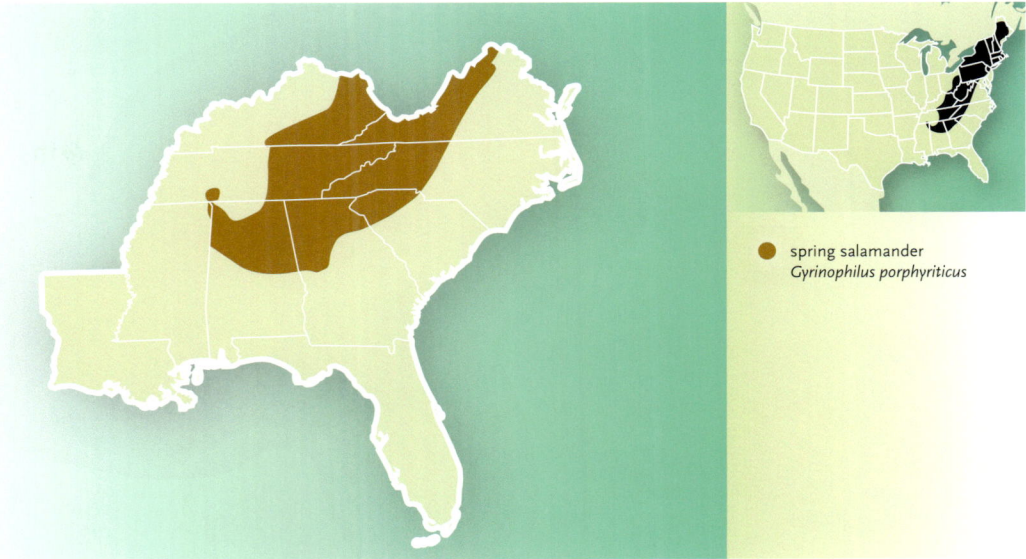

● spring salamander
Gyrinophilus porphyriticus

Spring salamander larvae (left and right) lack the light or dark spots found on the larvae of other salamanders in the region and may reach lengths of up to 6 inches before metamorphosing into adults. They are totally aquatic and have long, feathery external gills (right).

DISTRIBUTION Spring salamanders occur in the Southeast from western Virginia to northwestern South Carolina, northern Georgia and Alabama, and northeastern Mississippi; and in the eastern portions of Tennessee and Kentucky. Of the four subspecies, northern spring salamanders have the largest range. Kentucky spring salamanders are found in eastern Kentucky and southwestern Virginia; Blue Ridge spring salamanders occur in the Blue Ridge region of eastern Tennessee and western North Carolina; and the Carolina spring salamander ranges from southwestern North Carolina to east-central Alabama.

HABITAT Spring salamanders inhabit cool streams, springs, and caves in the Appalachian Mountains. They tend to avoid stream reaches with fish and are most abundant in headwater streams and seepage areas. Adults hide under rocks and in debris in streams during the daytime and prowl at night for prey, both in the water and on land. The larvae remain in gravel and under

Spring Salamander • 161

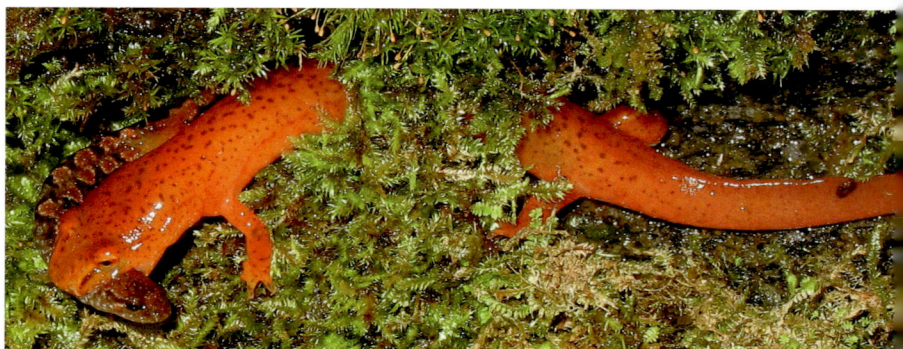

Spring salamanders have stout, robust bodies and a flattened tail. Older individuals like this one (top) often become darker in body color.

White flecking is sometimes evident on the lower jaw of spring salamanders (bottom left).

Spring salamanders are voracious predators of other stream salamanders, such as this Ocoee salamander (bottom right), as well as of a variety of invertebrates. Adults will even eat larvae and juveniles of their own species.

rocks until nightfall. Adults move into the surrounding hardwood forests during wet periods and can be found under logs and rocks. Larvae and adults are commonly found in caves when they are present in an area.

BEHAVIOR AND ACTIVITY Spring salamanders can be found in streams in every month of the year; during winter months, even at high elevations, they move around beneath rock cover and in crevices. They occasionally venture out from the water on rainy nights and can be found on roads and under logs and rocks in the surrounding forest. The larvae may occur in high densities, in part because the larval period is so long.

FOOD AND FEEDING Spring salamanders are voracious predators of other stream salamanders—including dusky, red, two-lined, and woodland salamanders—and can influence both population sizes and individual growth rates of prey species. Numerous invertebrates such as worms, isopods, spiders, centipedes, beetle larvae, slugs, and snails are also eaten. Head size determines the size of prey a larva or adult can eat. Adults eat larvae and juveniles of their own species as well.

REPRODUCTION Most of the reproductive activities take place in water, below the surface. Mating takes place in fall and spring, and perhaps in winter in the southern portion of the range. Females lay 16–106 eggs in late spring and summer in thin layers under rocks in water by turning upside down and using their arched back and tail to brace themselves. The eggs hatch in summer and fall. Lowland populations reach maturity a year earlier and at a smaller size (about 3 inches in total length) than high-elevation populations (about 5 inches in total length).

PREDATORS AND DEFENSE Northern watersnakes, garter snakes, and shrews are known predators, and predatory fish probably take these salamanders as well. Adults will tuck their head beneath their body and elevate and wave their tail when attacked by small mammals. Secretions from the tail are noxious to shrews.

Spring salamanders occasionally venture away from streams on rainy nights and travel into the surrounding forest.

CONSERVATION Spring salamanders are listed as Threatened in Mississippi. Siltation caused by deforestation, urbanization, agricultural activities, and construction above headwater streams can cause spring salamander populations to decline.

TAXONOMY This species was first described in 1827. Four subspecies are recognized: the northern spring salamander (*G. p. porphyriticus*); Blue Ridge spring salamander (*G. p. danielsi*), described in 1901; Carolina spring salamander (*G. p. dunni*), described in 1941; and Kentucky spring salamander (*G. p. duryi*), described in 1930.

Mud salamanders typi-
cally are stout bodied
and have a blunt head
and brown irises. The
body and tail are usually
red with well-defined,
round black spots.

How do you identify a mud salamander?

BODY SHAPE
Robust body; short,
thick tail

**NUMBER
OF COSTAL
GROOVES**
16 or 17

BODY COLOR
Reddish to brownish

PATTERN
Distinct round,
black spots on body
and tail

BODY SIZE
Metamorphs 1.6
inches; adults to
8 inches in total
length

Mud Salamander   *Pseudotriton montanus*

DESCRIPTION Mud salamanders are stout bodied and blunt headed. Body size varies geographically (total length reaches 8 inches in the eastern sub-species, *P. montanus montanus*), and females get larger than males. The body color ranges from bright red or pinkish to reddish brown depending on the subspecies, local variation, and age and size of the individual. Well-defined, round black spots are scattered on the body and tail, although spotting may be reduced or absent in some forms. The irises are brown. The midland mud salamander (*P. m. diastictus*) is more brightly colored than the other subspecies, and the rusty mud salamander (*P. m. floridanus*) lacks distinct spots on the body. The body length of *P. m. montanus* aver-ages about 1.5 times that of the other subspecies.

LARVAE AND JUVENILES The gilled, aquatic larvae are less than 0.5 inch long at hatching and average about 1.5 inches at metamorphosis. They are pale brown with dark spots on the back and dark markings on the sides that form streaks. The larval stage lasts more than a year; most individu-als metamorphose after 15–17 months. Juveniles resemble adults in color and spotting pattern.

SIMILAR SPECIES The mud salamander is most likely to be confused with the red salamander. The body color and ill-defined spotting patterns are highly variable in both species, but the red salamander has yellowish eyes, and the mud salamander has solid brown eyes. In areas where spring salamanders overlap with mud salamanders, the latter are recognizable in not having a distinctive line from the eye to the tip of the snout.

DISTRIBUTION All four subspecies of mud salamanders are found in the Southeast, and each southeastern state has at least one of them. Mud salamanders occur from New Jersey south to northern Florida and west through southern Mississippi to easternmost Louisiana. An isolated population is known from east-central Mississippi, but the species is absent from the rest of that state and from the western and northern halves of Alabama and from northern Georgia.

HABITAT Mud salamanders typically live in wet areas with muck and mud, particularly those with sphagnum moss and a decaying leaf layer in hardwood forests. The source of water is variable and includes seepage areas

The larvae of mud salamanders (left) are aquatic with distinct gills.

Mud salamanders typically live in wet areas with muck and mud (right), particularly those with sphagnum moss and a decaying leaf layer.

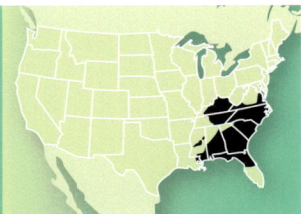

● mud salamander
Pseudotriton montanus

Mud salamanders may move short distances overland during the breeding season, but no well-defined seasonal activity patterns are apparent.

below bluffs, small streams, and floodplain swamps.

BEHAVIOR AND ACTIVITY Mud salamanders usually stay hidden, often in burrows in muddy, mucky sites; or beneath dead leaves, logs, and moss in wet habitats. They may move short distances over land during the breeding season, but no well-defined seasonal activity patterns are apparent.

FOOD AND FEEDING Adult mud salamanders are known to eat two-lined salamanders and worms. They presumably will eat other types of salamanders, small insects, and other invertebrates. The larvae eat aquatic invertebrates.

REPRODUCTION Mud salamanders mate in late summer or fall and probably lay eggs by December or January that hatch by February or March. Courtship has not been observed. Females in some South Carolina populations have been noted to carry 77–192 eggs, although lower numbers may be more common in other geographic regions. Actual clutch sizes above 30 eggs have not been found. The eggs are attached to the underside of moss, roots, or decaying vegetation in wet areas.

PREDATORS AND DEFENSE Northern watersnakes and eastern garter snakes eat mud salamanders; larger salamanders, ringneck snakes, shrews, and birds are other likely predators. When threatened, mud salamanders curl up so that the head is hidden beneath the body and wave the tail above the body. They produce skin secretions that are at least mildly toxic to some potential predators. The red body color may mimic that of other unpalatable red salamanders and thereby discourage attack by predators, such as birds, with color vision.

CONSERVATION No specific conservation issues have been targeted for this species, although local populations are likely to be affected by habitat degradation that affects stream systems and seepage areas.

TAXONOMY The mud salamander was first described in 1849. Four subspecies are recognized: the eastern mud salamander (*P. m. montanus*, described in 1849), midland mud salamander (*P. m. diastictus*, described in 1941), Gulf Coast mud salamander (*P. m. flavissimus*, described in 1856), and rusty mud salamander (*P. m. floridanus*, described in 1942).

Blue Ridge red salamanders like this one have little to no dark pigment on the last half of the tail.

Red Salamander

Pseudotriton ruber

DESCRIPTION Red salamanders are robust and have a thick tail. Females, which are larger than males, commonly exceed 6 inches in total length. The body and tail are red, but the shade ranges from deep or bright red to orangish, purplish, or brownish, depending on geographic variation and the size and age of the individual. Black spotting is present on most individuals; the spots are not distinct and often fuse together, especially on older animals, causing them to look purplish. The belly is pink to red and in adults usually has small black spots. The irises are yellow. Black-chinned red salamanders have heavy black pigment on the chin and lower lip; Blue Ridge red salamanders have little to no dark pigment on the last half of the tail and none on the chin and lower lip; and southern red salamanders are usually covered in white flecks, especially on the head and sides, in addition to the black spots.

LARVAE AND JUVENILES The hatchlings are about 0.5 inch in body length. The aquatic larvae have red gills; a pale brown body, sometimes with mottling on the back and sides; and a whitish belly. They metamorphose at about 2 inches in May to July after a growth period of 27–31 months. The larval stage in most populations lasts at least 1.5 years and can last more than 3 years in some. Juveniles resemble adults in color, but the belly lacks spots.

How do you identify a red salamander?

BODY SHAPE
Robust body with thick tail

NUMBER OF COSTAL GROOVES
16 or 17

BODY COLOR
Reddish to orangish or purplish

PATTERN
Black spots irregularly scattered over the entire body

BODY SIZE
Metamorphs 2 inches; adults to 7.1 inches in total length

Red Salamander • 167

SIMILAR SPECIES Red salamanders can usually be distinguished from mud salamanders, which may have a similar body color and pattern, by the yellow irises; mud salamanders have brown irises. In addition, the length of the snout from the eye to the nose tip is shorter in mud salamanders than in red salamanders. In areas of overlap with spring salamanders, red salamanders can be differentiated by the lack of a noticeable line from the eye to the tip of the snout.

DISTRIBUTION Red salamanders are found from southern New York, Pennsylvania, eastern Ohio, and southeastern Indiana through part of each southeastern state. They are absent from most of Louisiana and the Mississippi River floodplain of Mississippi, Tennessee, and Kentucky, as well as from most of the Atlantic Coastal Plain and the Florida peninsula.

HABITAT Red salamanders are associated with hardwood forests and with wet areas caused by natural seeps, stream systems, floodplains, and streams flowing through caves. The breeding and larval habitat is generally slow-moving, clear water with extensive decaying vegetation.

Hatchling red salamanders are about 0.5 inch in body length, and the aquatic larvae have red gills. The larval stage in most populations lasts at least 1.5 years and can last more than 3 years in some.

red salamander
Pseudotriton ruber

BEHAVIOR AND ACTIVITY Red salamanders stay beneath dead leaves, logs, and other ground litter or in burrows during the day, and may be active on the surface at night. They can be found in all seasons in and at the edge of water in springs, seeps, and small streams, as well as in terrestrial habitats. During spring and fall rains they may move long distances (more than 400 feet) over land between breeding sites and the terrestrial habitats where they spend most of the year. Individuals may be found under logs a long distance from water and are sometimes encountered on wet roads at night. During extreme cold periods red salamanders may go deeper into burrows or ground debris and become inactive.

Red salamanders are associated with hardwood forests and with wet areas, including those caused by natural seeps (top).

Some female red salamanders attach their eggs to the underside of a log or rock (bottom) and remain with them until they hatch.

FOOD AND FEEDING Red salamanders eat smaller salamanders as well as worms, insects, snails, spiders, and millipedes. In water, the larvae and adults eat aquatic insects and their larvae.

REPRODUCTION Red salamanders mate in summer or fall following courtship activity, and eggs are laid in fall or winter. The timing depends on such factors as geographic location and seasonal temperature and rainfall patterns. The number of eggs in a clutch ranges from as few as 29 to as many as 130. Some females attach the eggs to the underside of a log or rock and remain with them, sometimes creating and staying in a small depression beneath the eggs that fills with water. Some females lay their eggs deep in the banks of springs and small streams. The eggs hatch into aquatic larvae in 2–3 months.

PREDATORS AND DEFENSE Known predators include copperheads; watersnakes, garter snakes, ringneck snakes, shrews, raccoons, birds, and possibly larger salamanders are potential predators. When threatened, red salamanders exhibit elaborate body contortions by hiding the head and waving the tail slowly in the air. Their skin secretions are mildly toxic to some would-be predators. The red body color may mimic that of other red-colored salamanders that are mildly or strongly toxic (such as efts) and might therefore deter attacks by bird predators with good color vision.

The spots on some red salamanders are not distinct and often fuse together, especially on older animals like this one.

CONSERVATION The environmental health of this species overall appears to be satisfactory. On a local scale, populations suffer when timbering operations, mining, highway construction, urbanization, or other activities create excessive siltation, acid runoff, or other conditions that degrade the aquatic systems where these salamanders live. Loss of hardwood forests around streams is also detrimental.

TAXONOMY The red salamander was described in 1801. Four subspecies are recognized: the northern red salamander (*P. r. ruber*, described in 1801), black-chinned red salamander (*P. r. schencki*, described in 1912), Blue Ridge red salamander (*P. r. nitidus*, described in 1920), and southern red salamander (*P. r. vioscai*, described in 1928).

The male patch-nosed salamander (right) has two dark dorsolateral stripes and a yellow body whereas the female (left) is drab and lacks the dark stripes.

Patch-nosed Salamander

Urspelerpes brucei

DESCRIPTION Adult patch-nosed salamanders have a distinctive saffron yellow patch on the upper surface of the snout and a thin, distinct yellow line along the midline of the tail. The belly is uniformly yellow. Males have two distinct dark dorsolateral stripes and a yellow body. Females are drab and lack the dark stripes. An indistinct dark line runs posteriorly from in front of the eye to near the front limbs on each side. Reproductively active males have two prominent cirri and a mental gland. Adults are small—approximately the size of newly metamorphosed two-lined salamanders. Each rear foot has five toes, and the tail is as long as the body.

LARVAE AND JUVENILES The upper surface of the snout of larvae has a distinctive white patch. The body and tail are covered densely with small, brownish melanophores except on the snout and the white stripe on the tail.

SIMILAR SPECIES Patch-nosed salamanders resemble dwarf and two-lined salamanders. Dwarf salamanders have four toes on the rear feet, and the tail of both dwarf and two-lined salamanders is much longer than the body.

DISTRIBUTION Patch-nosed salamanders occur in one small area in the Appalachian foothills of northeastern Georgia (Stephens County) and northwestern South Carolina (Oconee County).

How do you identify a patch-nosed salamander?

BODY SHAPE
Slender

NUMBER OF COSTAL GROOVES
15

BODY COLOR
Yellow to yellowish brown

PATTERN
Males with two dark stripes on the back; females with no stripes

BODY SIZE
Adults to 2 inches in total length

● patch-nosed salamander
Urspelerpes brucei

HABITAT Little is known about this newly described salamander. Adults apparently prefer cover under rocks and in leaf litter on the moist banks of small first-order streams. The tiny larvae live in water in the sand-gravel interstices. Current information indicates that adults spend most of their time underground.

BEHAVIOR AND ACTIVITY Nothing is known about the behavior and seasonal activity patterns of patch-nosed salamanders.

FOOD AND FEEDING The structure of the mouth and tongue of adults suggested to the original describers of this species that they capture small insects and other invertebrate prey out of water.

REPRODUCTION All of the three adult females known were gravid, with enlarged ova visible through a translucent belly; two had 14 enlarged ova, and one had 6. Adult males collected between April 15 and May 29 had pronounced nasal cirri and enlarged mental glands, suggesting that the breeding season at least covers these two months.

PREDATORS AND DEFENSE No predators of this species have been described. Nothing is known about its defensive behavior.

CONSERVATION This species appears to be extremely rare and may occur in such small numbers as to be in danger of extinction.

TAXONOMY Patch-nosed salamanders were first described in 2009 by Carlos Camp and coauthors, making it the most recently described species in this book. The new salamander genus (*Urspelerpes*) is the first described in the United States since *Phaeognathus* in 1961.

SEASONAL WETLAND SALAMANDERS

A mature adult frosted flatwoods salamander of the uniform grayish brown phase. Note the short snout and robust body.

How do you identify a flatwoods salamander?

BODY SHAPE
Slender with small head

NUMBER OF COSTAL GROOVES
13–16

BODY COLOR
Brown to black

PATTERN
Frosted with reticulate or irregular specks

BODY SIZE
Metamorphs 2.75 inches; adults to 5 inches in total length

Flatwoods Salamanders

Frosted Flatwoods Salamander   *Ambystoma cingulatum*

Reticulated Flatwoods Salamander   *Ambystoma bishopi*

DESCRIPTION Frosted flatwoods salamander adults are brown to black with a reticulate pattern of gray to silver "frosting" on the back, head, and tail. The frosting may be irregular or may form a pattern that varies from narrow light lines to netlike. The frosting is concentrated on the lower sides of the body. The head is small and barely larger than the neck. The belly is dark gray with discrete light gray specks or white spots. Males in breeding condition have slightly swollen cloacal glands, but these do not get as large as in other mole salamanders. Reticulated flatwoods salamanders have narrow, orange-gray to silvery lines on a chocolate-black background; the belly has a salt-and-pepper pattern with tiny gray flecks.

LARVAE AND JUVENILES Larvae are aquatic and have bushy, reddish brown gills. The dorsal tail fins extend almost to the head. Young larvae have a yellow to gold stripe along the back with a black zone on each side. The

Frosted flatwoods salamander larvae have a prominent yellow to light tan stripe above the black stripe on the sides of the body and tail.

yellow stripe changes to tan on older larvae, which have two yellow to light tan stripes on each side of the body and tail; the stripe along the midbody is most prominent. The stripe on the side extends from the mouth and narrows as it passes along the tail. A dark stripe extends from the snout through the eye to the gills, and some older larvae have a black line along the mouth. The belly is cream to white, and the tail fin may have brown mottling. The period from hatching to metamorphosis into terrestrial juveniles is about 12 weeks. Juveniles may retain the tan stripes for nearly a year.

SIMILAR SPECIES Mole salamanders are short and chunky with a large head, and the body is solid gray or has light gray flecking. Spotted salamander juveniles are gray with fine black specks on the back but not on the tail. Juvenile marbled salamanders are brown to black with tiny gray to silver specks that may occur in a lichen-like pattern on the back.

DISTRIBUTION Flatwoods salamanders historically occurred in scattered locations from the central portion of the South Carolina Coastal Plain south to Alachua County, Florida, and west through the Florida panhandle to southern Alabama. Populations east of the Apalachicola River are frosted flatwoods salamanders; those west of the river are reticulated flatwoods salamanders.

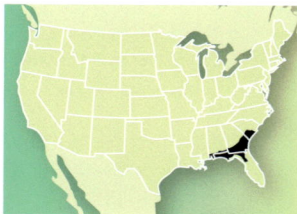

Metamorphic reticulated flatwoods salamanders acquire adult coloration and pattern as they grow and mature. The scattered lichen-like markings that cover the body of this metamorph will fade with age.

- ● frosted flatwoods salamander
  Ambystoma cingulatum

- ● reticulated flatwoods salamander
  Ambystoma bishopi

An adult frosted flatwoods salamander with a somewhat typical pattern of lichen-like markings on a nearly black body.

HABITAT Adults of both species remain underground most of the year, presumably in rodent burrows and root tunnels in slash pine flatlands and longleaf pine forests. Where the juveniles go in the upland habitat after they leave the breeding ponds is unknown. Breeding sites include cypress and blackgum swamps, roadside ditches, ephemeral pools, and shallow borrow pits.

An adult reticulated flatwoods salamander with profuse silvery frosting and narrow reticulations across the body and tail.

BEHAVIOR AND ACTIVITY Adults emerge from upland refugia on rainy nights in October and November, mate and lay eggs in the breeding pond, and migrate back to their burrows during rains in December and January. Eggs hatch from November to January, and the larvae remain in the pond until March or April. The larvae stay hidden among aquatic plants in shallow water during the day and forage in the open water column at night. After metamorphosis, juveniles migrate into the surrounding uplands and presumably remain underground until they reach maturity.

FOOD AND FEEDING Adults and juveniles eat earthworms and small slugs. The aquatic larvae eat small invertebrates and zooplankton.

REPRODUCTION Courtship and mating take place on land as males and females are approaching breeding ponds. Females lay eggs in small groups of 1–4 eggs in the breeding pond basin, usually under logs and other surface objects that will become flooded during winter rains; or they attach the eggs to vegetation in standing water. Clutch size ranges from about 100 to 225 eggs.

PREDATORS AND DEFENSE Specific predators are unknown, but aquatic invertebrates such as dragonfly nymphs, predaceous diving beetle larvae, and giant water bugs probably eat the larvae and even adults that remain in the aquatic habitat during warm weather. Large larvae may be cannibalistic. The dark pattern of brown mottling on the tail is likely used to distract predators so that they attack the tail instead of the head and body. Adults will coil the body and wrap the tail around the head when threatened.

The dark pattern of brown mottling on the tail is likely used to distract predators so that they attack the tail instead of the head and body.

CONSERVATION Frosted flatwoods salamanders are listed as Threatened by the U.S. Fish and Wildlife Service as a result of habitat loss, disease, the small number of currently known locations, and the apparent dramatic decline in numbers at known breeding sites. Alabama, Florida, Georgia, and South Carolina also protect this species either as a Species of Concern or by listing it as Endangered. The reticulated flatwoods salamander is protected under the Endangered Species Act as Endangered because it faces the same threats faced by the frosted flatwoods salamander.

TAXONOMY The frosted flatwoods salamander was first described in 1868. The reticulated flatwoods salamander was described as a subspecies in 1950 and recognized as a full species in 2006. No subspecies are recognized for either species.

Adult Jefferson salamanders are typically uniformly colored without markings, but many retain remnants of silvery flecking and spots.

Jefferson Salamander   *Ambystoma jeffersonianum*

DESCRIPTION Jefferson salamander adults are larger and have a broader head and more elongated snout than other mole salamanders. They are light to dark bluish gray with silver flecking on the lower sides of the body, tail, and limbs. Some adults lack the flecking altogether, and some have silver flecking on the back.

LARVAE AND JUVENILES Larvae are olive green to gray with faded yellow markings on the upper tail fin and on the sides of the head and neck. The belly is silver to white and unmarked. Metamorphs are uniformly gray with faded yellow spots on the sides and an unmarked belly.

Jefferson salamander larvae (top) have large gills and are olive green to gray with faded yellow markings on the upper fin and on the sides of the head and neck.

Newly emerged metamorph Jefferson salamanders (bottom) are typically thin bodied with a large head.

How do you identify a Jefferson salamander?

BODY SHAPE
Long and robust; broad head

NUMBER OF COSTAL GROOVES
12 or 13

BODY COLOR
Bluish gray

PATTERN
Uniform or with fine silver flecking concentrated on the sides

BODY SIZE
Metamorphs 2 inches; adults to 7 inches in total length

Adult Jefferson salamanders are larger and have a broader head and more elongated snout than other mole salamanders.

The yellow spotting of metamorphs changes to silvery spots and flecking in immature salamanders, but these fade with age.

SIMILAR SPECIES Marbled salamander juveniles are brown to black with tiny gray to silver specks that may occur as a lichen-like pattern or as paired faded yellowish spots on the back. Spotted salamander juveniles are olive to gray with fine gray specks on the back but not on the tail. Immature small-mouthed salamanders are brownish gray to black with clusters of light specks or a lichen-like pattern mostly concentrated on the sides. Newly metamorphosed juvenile streamside salamanders are uniformly brown to brownish gray with little or no pattern.

DISTRIBUTION Jefferson salamanders occupy high elevations in the Appalachian Mountains and northern latitudes from southwestern Virginia and central Kentucky northeastward to Ontario, Canada, and New Hampshire.

HABITAT These are primarily salamanders of well-drained hardwood forests in montane regions. Adults and juveniles occupy rodent burrows and root tunnels.

BEHAVIOR AND ACTIVITY Adults emerge from underground refugia in late fall before the ground freezes and in winter as soon as the snow melts and the ground thaws, and sometimes before that. They migrate long distances—up to a mile—to reach their woodland breeding pools. Breeding

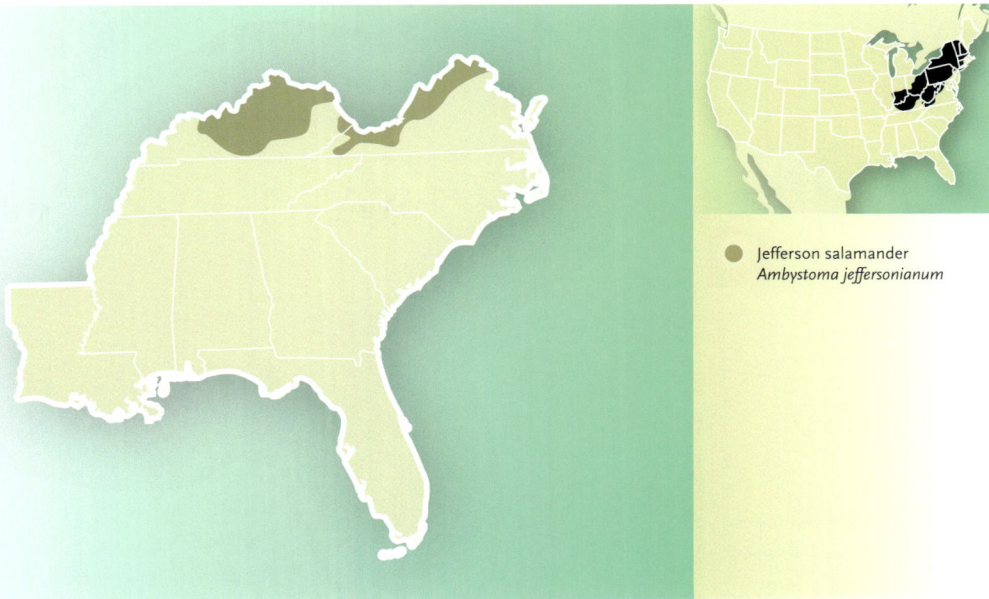

sites may be ephemeral or permanent but are usually devoid of fish, which are major predators. Adults leave the ponds 3–4 weeks after breeding and laying eggs and migrate back to their underground retreats in the forest. They are seldom seen outside the breeding period because they are usually underground.

FOOD AND FEEDING Larval Jefferson salamanders eat a wide variety of aquatic invertebrates and zooplankton. They will also eat the larvae of other salamander species, as well as each other. Adults are thought to eat invertebrates such as earthworms while in terrestrial burrows.

REPRODUCTION Males usually arrive at the breeding pond before females and lay numerous spermatophores on the pool bottom. Once females arrive, a male will entice a female by behavioral and chemical communication to walk over and pick up a spermatophore with her cloaca. Fertilization of the 140–200 eggs is internal. Eggs are laid 1–2 days later in several gelatinous masses that the female attaches to twigs or other vegetation. The larval period lasts 2–4 months during the summer. Larvae remain hidden under leaves during the day and forage in the open water column at night. Metamorphs move as far as a third of a mile into forest habitat on land and remain underground until they mature 2–3 years later.

Jefferson salamanders lay their eggs in clusters attached to stems and grasses in water.

Jefferson salamanders are found primarily in well-drained hardwood forests. Timber harvest and loss of forest habitat threaten many populations in Kentucky and Virginia.

PREDATORS AND DEFENSE Striped skunks, raccoons, and shrews are known predators of adults; owls and snakes are suspected predators. Adults elevate the tail, coil the body, hide the head under the tail, and produce a noxious secretion from tail glands that deters potential predators. The larvae are eaten by aquatic invertebrate predators and sometimes by other Jefferson salamanders. The tail fins of larvae become dark in the presence of aquatic invertebrate predators and presumably serve as a decoy that attracts the predator to the tail rather than the head.

CONSERVATION Timber harvest and loss of forest habitat threaten many populations in Kentucky and Virginia. Acid precipitation that lowers pH in the breeding ponds is another human-made source of mortality. Maintaining mature forests and several breeding ponds in the landscape will be necessary for long-term survival of Jefferson salamander populations.

TAXONOMY The Jefferson salamander was first described in 1827. No subspecies are recognized, although in the northern portion of its range this species hybridizes with blue-spotted salamanders to form triploid offspring that are all females.

An adult mole salamander showing the chunky head and body and typical color pattern in the Southeast

How do you identify a mole salamander?

Mole Salamander

Ambystoma talpoideum

DESCRIPTION Adults and juveniles are chunky with a large, wide head and large feet; the tail is short for their size. They are light brown, various shades of gray, or nearly black with light gray specks on the back, sides, and tail. Some individuals lack the speckling altogether. The belly of older individuals is uniformly light gray to cream; recently metamorphosed individuals have a belly stripe.

LARVAE AND JUVENILES Larvae are gray to olive brown with two creamy to yellowish stripes on the body that fade to blotches on the tail. The upper stripe may or may not be conspicuous, and the lower one may fade into the belly but is usually distinct. A narrow dark gray to black line, often bordered by parallel pale yellowish lines, is usually present along the middle

BODY SHAPE
Chunky with a big head; large legs; short tail

NUMBER OF COSTAL GROOVES
10 or 11

BODY COLOR
Brown, gray, or black

PATTERN
None, or small gray flecks on the back and tail

BODY SIZE
Metamorphs 2.2 inches; adults to 4.5 inches in total length

Mole salamander larvae are gray to olive brown with two cream to yellowish stripes on the body that fade into blotches on the tail. The belly has a dark stripe along the midline.

The silver flecking on this immature mole salamander will largely disappear as it matures.

The abundant silver flecking on the tail and sides of this adult mole salamander illustrate individual variation in pattern.

of the belly. Juveniles are often greenish brown and may retain the yellowish stripe on the sides but may also show some gray flecking.

SIMILAR SPECIES Adult marbled salamanders are also chunky but have white to cream crossbars on the back. Their juveniles are brown to black with tiny gray to silver specks that may occur as a lichen-like pattern or as paired faded, yellowish spots on the back. Spotted salamander juveniles are either olive to gray with fine gray specks on the back but not on the tail, or have the spots typical of adults. Immature small-mouthed salamanders are brownish gray to black with clusters of light specks or a lichen-like pattern mostly concentrated on the sides. Newly metamorphosed juvenile streamside salamanders are brown to brownish gray with little or no pattern.

DISTRIBUTION This southeastern species occurs from south-central Virginia south to north-central Florida; west through Louisiana; and north to northeastern Arkansas and southeastern Missouri, western Tennessee, and western Kentucky. Many isolated populations occur in Virginia, the Carolinas, Georgia, Alabama, and Tennessee.

HABITAT Mole salamanders occupy bottomland hardwood and floodplain forests near gum and cypress swamps and Carolina bays in the Coastal Plain. They also occur in hardwood forest and mixed hardwoods and pine forest around vernal pools, borrow pits, ditches, and other ponds outside the Coastal Plain. Breeding ponds usually lack fish and dry up in summer.

BEHAVIOR AND ACTIVITY Adults remain underground in refugia or beneath pine straw or other ground litter when not in breeding ponds. Adults in the southern portion of the range migrate from the upland forests to breeding ponds from November to January during nighttime rains. Those in northern locations move to ponds during the late winter months. Mating occurs in the water immediately after arrival. Adults leave the ponds within several days to a week and migrate back to their forest refugia. Populations of neotenic individuals remain in the pond year round. If the pond dries, they will absorb their gills and move into forest refugia.

FOOD AND FEEDING Aquatic larvae consume a wide variety of small-bodied invertebrate prey, tadpoles, and other salamander larvae, including their own species. Adults probably eat earthworms, leeches, and other invertebrates.

REPRODUCTION Terrestrial adults without gills mate and lay eggs in October through March, with the timing depending on weather conditions and the geographic area. Populations of

Some individuals in southeastern populations remain in the water for long periods and acquire the ability to reproduce before metamorphosis. These paedomorphic mole salamanders will lose their gills and emerge from their pond when their habitat dries up.

mole salamander
Ambystoma talpoideum

Individual color variants resulting from genetic anomalies, such as this larval mole salamander lacking black pigment except in the eyes, are likely to occur in many populations but at low frequencies.

Female mole salamanders in some areas attach a clutch of eggs to twigs and branches while females in other areas deposit single eggs on the substrate of the breeding pond.

neotenic gilled adults mate as early as September and as late as November. Some populations in permanent ponds have both neotenic and terrestrial breeding adults. Males deposit about 15 spermatophores on the substrate and entice females to pick up 1 or more of them with the cloaca. Females in Gulf Coast populations deposit 4–99 eggs in small clusters on twigs; those in Atlantic Coast states lay them singly on the pond bottom. However, total clutch size may be as high as 200–700 per female. Neotenic females are usually smaller than terrestrial females and lay fewer eggs. The gilled adult population will produce mature individuals by the following September. Larvae from adults without gills metamorphose in May to September but may not mature for several more months.

PREDATORS AND DEFENSE Fish, especially bluegills, are the primary predators of eggs and larvae, which hide in the leaf litter when fish are present. Watersnakes (southern banded, plain-bellied, and eastern green), black swamp snakes, cottonmouths, black racers, mud snakes, rainbow snakes, ribbon snakes, garter snakes, and wading birds eat adults. When threatened, adults use a head-down posture, become immobile, bite, head butt, flip the body, or wave the tail. Glands in the tail and the paratoid gland behind the head produce noxious secretions that may deter predators.

CONSERVATION Mole salamanders are listed as a Species of Special Concern in North Carolina and as a Species in Need of Management in Tennessee. Primary threats include urban sprawl, agricultural practices, deforestation, and changes in landscape hydrology.

TAXONOMY The mole salamander was first described in 1838. No subspecies are recognized.

How do you identify a small-mouthed salamander?

BODY SHAPE
Chunky with a small, short head and mouth

NUMBER OF COSTAL GROOVES
13–15

BODY COLOR
Gray, brown, or black

PATTERN
Pale, lichen-like

BODY SIZE
Metamorphs 2.2 inches; adults to 6 inches in total length

Small-mouthed Salamander

Ambystoma texanum

DESCRIPTION The noticeably small head and mouth distinguish this species from other species of mole salamanders. Adults are shiny gray to grayish black with light gray, lichen-like speckling on the back and sides that is often concentrated on the sides. Some individuals are almost uniformly black. The belly is gray with light speckling or black.

LARVAE AND JUVENILES Hatchlings and small juveniles are brownish to olive green with three to six pairs of yellowish olive green patches along the middle of the back. Older larvae are light brown with inconspicuous spots. Metamorphs are brown to gray and unmarked; the lichen pattern appears a few weeks after metamorphosis.

Older larvae of small-mouthed salamanders are light brown with inconspicuous spots.

SIMILAR SPECIES Other mole salamanders have a normal-sized head. Marbled salamander juveniles have a lichen-like pattern of tiny gray to silver specks on a brown to black body or a series of paired faded, yellowish spots on the back. Spotted salamander juveniles are olive to gray with fine gray specks on the back but not on the tail. Larval mole salamanders are gray with two creamy to yellowish stripes on the body that fade to blotches on the tail. Newly metamorphosed juvenile streamside salamanders are brown to brownish gray with little to no pattern.

DISTRIBUTION Small-mouthed salamanders occur in western Alabama and Tennessee and in the states bordering the Mississippi River southward to the Gulf coast of Louisiana.

HABITAT Adults and juveniles occupy bottomland hardwood forests along rivers and streams, and hardwood forests around ephemeral ponds and swamps. Juveniles hide under leaf litter and objects such as logs and debris.

BEHAVIOR AND ACTIVITY Adults live and forage beneath the forest floor. They emerge en masse on rainy nights in late winter to early spring and migrate to breeding ponds. Reproduction occurs in a wide variety of shallow, temporary wetlands such as woodland pools, ditches, borrow pits, flooded fields, and swamps. Juveniles spend most of their time under the leaf litter.

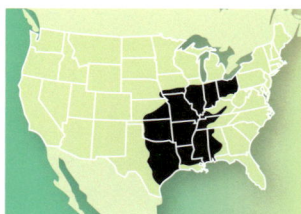

The pattern on the back of some small-mouthed salamanders is distinct from the pattern on the sides.

small-mouthed salamander
Ambystoma texanum

Adult small-mouthed salamanders may be uniformly brown, although the gray phase is more common.

FOOD AND FEEDING Adults and juveniles eat a wide variety of prey including earthworms, centipedes, moths, crickets, and beetles. Larvae eat small invertebrates such as isopods and copepods and may be cannibalistic. The size of the salamander's mouth dictates the size of the prey it can swallow.

REPRODUCTION Males usually arrive at breeding ponds before females during winter or early spring and place spermatophores on leaves, twigs, and other bottom features. Females arrive shortly afterward. As each is courted by a male, she uses her cloaca to pick up a sperm packet sitting atop a spermatophore; fertilization is internal. One female may pick up several sperm packets, not all necessarily from the same male. Females deposit 250–800 eggs singly, in loose clusters, or in gelatinous clusters of 4–30 eggs each. Hatching occurs in later winter. The larvae remain in the pool for 2–4 months, then emerge during rain as metamorphs.

Small-mouthed salamander females lay their eggs in small clusters attached to twigs and stems in the water.

PREDATORS AND DEFENSE Known predators of larvae include tiger salamanders, predaceous insects, garter snakes, and watersnakes. When threatened, adults lower the head, curl the body, and wave the tail in an attempt to disperse noxious gland secretions that may discourage a predator. Larvae also use concealment and rapid swimming to escape.

Adults are shiny gray to grayish black with light gray, lichen-like speckling on the back and sides that is often concentrated on the sides.

CONSERVATION Urbanization and agriculture threaten wetland breeding sites of small-mouthed salamanders and the adjacent forest cover where they spend most of their lives. Many adults and metamorphs are killed on roads during migrations to and from breeding ponds. No southeastern state wildlife department classifies this species as Endangered or Threatened.

TAXONOMY The small-mouthed salamander was described initially in 1855. No subspecies are recognized.

Tiger salamanders are the largest terrestrial salamanders in the Southeast. Their big head allows them to take large prey.

How do you identify a tiger salamander?

Tiger Salamander

Ambystoma tigrinum

DESCRIPTION Adult tiger salamanders are the largest mole salamanders in the Southeast—up to 13 inches in total length. The body is robust, and the head is large. The tail of males is compressed side to side and is larger than that of females. The body is bluish gray to brown to nearly black. The back, sides, and tail have olive to yellowish brown, irregularly shaped spots. The belly is olive yellow to cream with some faded dark pigment in a marbled pattern.

LARVAE AND JUVENILES Larvae are olive to gray with paired dark blotches on the body and tail and a white belly. Their toes are wedge shaped; those of all other mole salamander larvae are round. Juveniles are olive gray to dark brown with at least some irregular spots on the body.

BODY SHAPE
Robust; large head; long tail

NUMBER OF COSTAL GROOVES
11–14

BODY COLOR
Olive gray to black or brown

PATTERN
Yellowish, irregular spots on body, head, and tail

BODY SIZE
Metamorphs 2–4 inches; adults to 12 inches in total length

Larval tiger salamanders are olive to gray with pairs of dark blotches on the body and tail. Each toe is wedge shaped.

SIMILAR SPECIES Marbled salamander juveniles are brown to black with tiny gray to silver specks that may occur as a lichen-like pattern or as pairs of faded yellowish spots on the back. Spotted salamander juveniles are olive to gray with fine gray specks on the back but not on the tail. Juvenile Jefferson salamanders are uniformly dark gray, occasionally with brownish yellow specks on the sides. Immature small-mouthed salamanders are brownish gray to black with clusters of light specks or a lichen-like pattern mostly concentrated on the sides. Newly metamorphosed juvenile streamside salamanders are brown to brownish gray with little or no pattern.

Three tiger salamander juveniles illustrate some of the range of variation in body pattern at this early stage.

DISTRIBUTION Tiger salamanders are found throughout much of the Southeast from north-central Florida north through the Coastal Plain to southeastern Virginia, and northwest through Alabama and Tennessee and western Kentucky. Isolated populations occur in Arkansas, Tennessee, and Virginia. The species also is found throughout much of the Midwest and as far west as Texas.

HABITAT Adults in the Southeast live almost anywhere there is forest cover and soil in which they can burrow; they primarily occupy hardwood forest habitats but may also occur in longleaf pine forests, mixed pine

The spotted pattern of this adult tiger salamander is found in some southeastern populations.

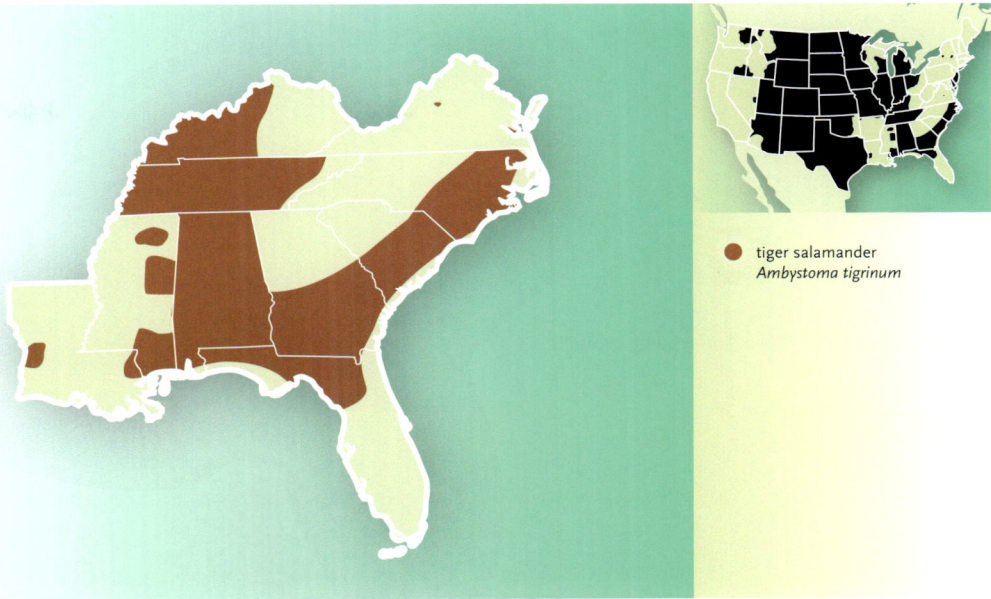

● tiger salamander
  Ambystoma tigrinum

and hardwood forests, and bottomland forests. Breeding takes place in seasonal to permanent ponds that lack fish, as well as in roadside ditches, artificially created wildlife ponds, and slow-moving streams.

BEHAVIOR AND ACTIVITY Unlike some other mole salamanders, tiger salamanders dig their own burrows in the soil. They remain underground—and are seldom seen—between breeding seasons, and migrate to breeding ponds during warm rains in winter months. Breeding ponds are usually less than 300 yards from their upland refugia, although such distances vary depending on habitat type and other aspects of the landscape.

Tiger salamanders will eat any animal they can capture and subdue.

FOOD AND FEEDING The larvae consume zooplankton, a wide variety of aquatic invertebrates, worms, leeches, crayfish, tadpoles, other salamander larvae, and sometimes each other. Individuals in some populations become larger than the other tiger salamanders, grow larger teeth, and become cannibals. Adults do not feed in the breeding season but in their upland habitats eat any prey they can get in their mouth, including a wide variety of invertebrates and even mice.

REPRODUCTION Males arrive as early as late December and mate in the water with females when they arrive 2–4 weeks later. Mating is sometimes under ice on the pond's surface. The males deposit spermatophores on the

Developing tiger salamander eggs are visible within the egg capsule and jelly-like egg mass.

pond substrate that females pick up with the cloaca and use to fertilize their eggs. Gelatinous masses of 5–122 eggs are attached to twigs. Adults leave the pond after mating and return to their burrows. The larvae hatch several weeks after the eggs are laid and reach metamorph size in 10 weeks to 6 months.

PREDATORS AND DEFENSE Known aquatic predators include predaceous diving beetles, dragonfly nymphs, other tiger salamanders, southern banded watersnakes, mud snakes, and wading birds. Terrestrial predators include shrews, middle-sized mammals such as raccoons, and snakes. Larvae hide during the day in vegetation and under leaf litter at the bottom of the pool and forage in the open water column on moonless nights. Adults produce a sticky, toxic secretion in glands in their tail that repels and may even kill predators.

CONSERVATION Eastern tiger salamanders are protected as Endangered in Virginia and Threatened in North Carolina, and are considered a Species of Moderate Conservation Concern in Alabama. They are not protected in any other southeastern state. The primary threat is loss of both upland habitat and pond breeding sites to urban sprawl. Populations at high elevations are affected by acid precipitation. Mortality can be extremely high on roads between breeding sites and upland habitats. Several diseases are known to occur in some populations in the Southeast, but their impacts are unknown. Upland and breeding habitats in areas where populations are in danger of extirpation need proper management if the tiger salamanders are to be preserved.

TAXONOMY The tiger salamander was first described in 1825. The several subspecies recognized in western North America have been grouped within the barred tiger salamander species (*Ambystoma malvortium*) complex, leaving the eastern tiger salamander without subspecies.

This subadult Mabee's salamander has profuse silver flecking and spotting on the back that will fade as it ages.

Mabee's Salamander

Ambystoma mabeei

DESCRIPTION Adults are medium-sized mole salamanders with a small head and small round patches of light gray to silver specks all over the black body, head, and tail. The markings run together and concentrate along the sides of the body and are relatively sparse on the back, but the individual pattern is highly variable. The belly is dark brown to gray. As in other species of mole salamanders, males in breeding condition have swollen cloacal glands; the cloacal area of females is flat.

LARVAE AND JUVENILES The aquatic larvae have bushy gills, dorsal fins that extend almost to the front limbs, and a yellowish stripe bordered above and below with black along each side. Older larvae have a second yellow stripe below the lower black zone, and both stripes may be broken or appear blotched. The belly is cream-colored, and the tail fin has dark black mottling. The specks on juveniles are small and not clustered as in older adults, and are not concentrated along the sides of the body. Larvae transform into terrestrial juveniles after 2–9 months, depending on when their woodland ponds fill with water and subsequently dry.

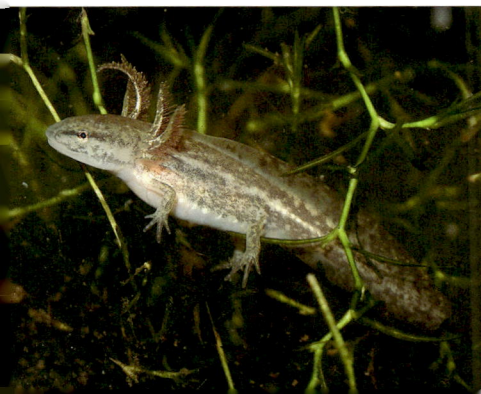

Mabee's salamander larvae are gray to olive brown with two cream to yellowish stripes on the body that fade into blotches on the tail.

How do you identify a Mabee's salamander?

BODY SHAPE
Slender; small head

NUMBER OF COSTAL GROOVES
13

BODY COLOR
Black to dark brown

PATTERN
Round patches to scattered patches of gray flecks

BODY SIZE
Metamorphs 2.2 inches; adults to 4.5 inches in total length

SIMILAR SPECIES Mole salamanders are gray and chunky with a large head. Spotted salamanders are black with yellow spots, and the juveniles are gray with fine black specks on the back but not on the tail. Juvenile marbled salamanders are brown to black with tiny gray to silver specks that change into a lichen-like pattern within a month after metamorphosis.

DISTRIBUTION Mabee's salamanders occur in the Atlantic Coastal Plain from southeastern Virginia to the southern portion of the Savannah River in South Carolina. No records are known in the Piedmont, north of the Fall Line.

HABITAT Adults and juveniles occupy animal burrows and root tunnels in open fields, pine forests, and hardwood forests when not at the breeding pond. They may be found under all manner of surface cover objects during wet periods. Breeding sites include vernal pools, ditches, small fishless ponds, and Carolina bays.

Mabee's salamander metamorphs are uniformly gray to brown and develop silvery flecking.

BEHAVIOR AND ACTIVITY Adults and juveniles may be active on the surface during rains but otherwise remain underground. Surface activity increases as the breeding season approaches. In January, February, and March they migrate from their terrestrial refugia to wetlands up to half a mile away where they mate and lay eggs. Adults remain in the pond for several days to

Mabee's salamander
Ambystoma mabeei

weeks after breeding and then migrate back to their upland refugia, usually on a rainy night.

Adult Mabee's salamanders have a uniformly gray back with abundant silvery flecking and markings on the sides.

FOOD AND FEEDING The larvae feed on zooplankton and small invertebrates. Mouth size determines the size and type of prey that they can eat. Adults are known to eat earthworms in upland habitats.

REPRODUCTION The courtship behavior of this salamander has not been published. Females attach eggs singly or in strings of 2–6 to leaves and other material on the bottom of the pond. Individual female clutch size is unknown.

PREDATORS AND DEFENSE The larvae are eaten by tiger salamander larvae, lesser sirens, fish, dragonfly nymphs, and predaceous water beetle larvae. Predators of adults have not been observed but likely include shrews and snakes that find them underground. When threatened, adults become immobile, hide the head behind the tail, curl and wave the tail to distract the predator, and tuck the head down to the substrate to minimize exposure.

CONSERVATION Mabee's salamanders are listed as a Threatened species by the state of Virginia. Threats include habitat loss, road mortality, and alteration of pond hydrology resulting from ditching for urban developments.

TAXONOMY Mabee's salamander was first described in 1928. No subspecies are recognized.

This adult male marbled salamander has bright crossbands on his body and tail and abundant white pigment on his head. Individual salamanders can be recognized by their pattern.

How do you identify a marbled salamander?

BODY SHAPE
Chunky

BODY COLOR
Black

PATTERN
White or light gray crossbands across the head, back, and tail

BODY SIZE
Metamorphs 1.8–2 inches; adults to 4.5 inches in total length

Marbled Salamander   *Ambystoma opacum*

DESCRIPTION The fully terrestrial adults are medium-sized and have a chunky build. The black ground color has an overlay of contrasting white or light gray crossbands across the head, back, and tail. The crossbands may appear as stripes on some individuals, and rare individuals are entirely black. The belly is uniformly black. The crossbands are silvery white on males and silvery gray on females. Males in breeding condition also can be distinguished from females by their swollen cloacal glands.

LARVAE AND JUVENILES The aquatic larvae have bushy gills, a dorsal fin that extends almost to the front limbs, and are drab brown or black with a series of light spots that form a line on the sides. Larvae transform into terrestrial juveniles after 2–9 months, the period depending on geographic location and when their woodland ponds fill with water and subsequently dry.

SIMILAR SPECIES Spotted salamander juveniles are dull olive green to dark gray with fine gray specks on the back but not on the tail. Mole salamanders are short and chunky with a large head and light gray or no flecking. Juvenile Jefferson salamanders are uniformly dark gray, occasionally with brownish yellow specks on the sides. Streamside salamander juveniles are uniformly brown to brownish gray.

DISTRIBUTION Marbled salamanders occur in much of the eastern United States from eastern Texas and Oklahoma northeast through Illinois and Indiana to southern New Hampshire and central Massachusetts, and south to northern Florida.

HABITAT Mating takes place at night in the open on the surface, and females lay eggs underneath cover objects in the dry basins of temporary, fish-free wetlands such as upland hardwood "swamp forests," bottomland hardwood pools, quarries, vernal ponds, Carolina bays, and floodplain pools. Outside the autumn breeding season terrestrial adults and juveniles live underneath leaf litter and in small mammal burrows in mature deciduous forest, mixed hardwoods, pine stands, floodplains, and uplands.

BEHAVIOR AND ACTIVITY Marbled salamanders may be inactive for prolonged periods in the summer when the weather is hot and dry. On rainy nights from September through November adults migrate to the dry ba-

Marbled salamander larvae (below left) closely resemble spotted salamander larvae in color and pattern but have black pigment on their throat and are usually larger.

Metamorphic marbled salamanders (below right) develop a highly variable amount of frosty flecking that becomes crossbars and other patterns in adults.

marbled salamander
Ambystoma opacum

Marbled salamanders (top) show considerable variation. This individual has two stripes down its back rather than the crossband pattern typical of most adults.

Females (bottom) lay their eggs in depressions around the margin of a breeding pond under leaves, moss, or logs and remain with them until the pool fills with water.

sins of temporary wetlands; males typically arrive at breeding sites before females. The adults return to the wooded habitat surrounding the wetland on rainy nights between late fall and winter.

FOOD AND FEEDING Adults feed on terrestrial invertebrates such as millipedes, centipedes, spiders, insects, snails, arthropods, worms, and mollusks. The larvae eat small aquatic invertebrates, including a variety of crustaceans such as copepods, amphipods, and isopods, and smaller salamander larvae.

REPRODUCTION Marbled salamanders are one of only two species of mole salamanders that breed on land, and they are the only member of the genus *Ambystoma* in which the female parent guards the developing embryos. Following courtship, females lay from 30 to nearly 200 single eggs in nests constructed underneath logs, vegetation, or leaf litter in the dried beds of ponds, the margin of reduced ponds, or dry floodplain pools, typically in a spot likely to be flooded by winter rains. Most females nest singly, although two to seven females have been observed nesting communally under the same cover object. Females typically stay with their nests to guard the eggs until the pond begins to fill and the larvae start to hatch; however, nests without attending females are not unusual.

PREDATORS AND DEFENSE The most common known predators of adults are eastern garter snakes, short-tailed shrews, raccoons, and opossums. The larvae are eaten by southern banded watersnakes, eastern green watersnakes, and black swamp snakes. Adults have granular glands on the head that produce noxious secretions known to deter attacks by shrews and common ribbon snakes.

CONSERVATION Loss of bottomland hardwoods and their associated wetland habitats poses the greatest risk to this species. Small, isolated wetlands are the most valuable habitats for maintaining high amphibian biodiversity, but they are also most at risk because current regulations generally do not protect them. Marbled salamanders are listed as Threatened in Massachusetts and Michigan, and as Protected in New Jersey.

TAXONOMY The marbled salamander was first described in 1807. No subspecies are recognized.

Adult spotted salamanders typically have yellow spotting on the head, body, and tail on a dark gray body.

Spotted Salamander

Ambystoma maculatum

DESCRIPTION Spotted salamanders are large and robust with a black body and distinct yellow or orange spots on the body, head, and tail. The belly is uniformly dark gray.

LARVAE AND JUVENILES The larvae are greenish to gray above with few markings and have an unmarked white belly. At metamorphosis, the robust juveniles have a thin-lined, reticulate pattern on a gray background or are uniformly gray with no markings. The spots appear within several months.

BODY SHAPE
Robust body; rounded head

NUMBER OF COSTAL GROOVES
11–13

BODY COLOR
Black

PATTERN
Distinct yellow spots

BODY SIZE
Metamorphs 2 inches; adults to 9 inches in total length

Typical spotted salamander larvae are brown to gray with bushy gills and yellow spotting on each side. They differ from marbled salamander larvae in lacking black pigment on the throat.

Metamorphic spotted salamanders are uniformly gray with few markings. The dark remnant of the gill bud on the back of the head indicates that this little salamander is a recent metamorph.

Adult spotted salamanders in some populations have orange spots, often on the head, in addition to the yellow spots on the body and tail.

SIMILAR SPECIES Larval marbled salamanders have scattered specks of dark pigmentation on the throat. Juvenile marbled salamanders are brown to black with tiny gray to silver specks that may occur as a lichen-like pattern on the back. Small-mouthed salamander adults are brownish gray to black with clusters of light specks or a lichen-like pattern mostly concentrated on the sides. Mole salamanders are short and chunky with a large head and are nearly uniform in color except for a scattered pattern of light gray flecks. Juvenile Jefferson salamanders are uniformly dark gray, occasionally with brownish yellow specks on the sides.

DISTRIBUTION Spotted salamanders occur widely in the eastern United States from Canada south through Virginia and the Carolinas to the Gulf of Mexico in Alabama and Mississippi, and westward to eastern Texas and southeastern Missouri. They are absent from the Coastal Plain of North Carolina, southeastern Virginia, southern Georgia, and most of Florida.

HABITAT Spotted salamanders live primarily in hardwood and mixed deciduous forest habitats but may be found in a wide array of habitat types, including mesic floodplains, if forest cover is present. They breed in shallow fresh water, often in isolated and ephemeral wetlands. The combination of wetlands and upland terrestrial habitats is essential to the survival of this salamander.

BEHAVIOR AND ACTIVITY Late-winter rains bring adults from forest refugia to shallow breeding ponds. Eggs hatch in 1–3 weeks, and the larval period lasts about 3 months depending on how elevation, latitude, and canopy cover affect water temperatures. Post-metamorphic animals migrate away from wetlands in mid-to-late summer. Adults and juveniles hibernate in subterranean burrows in the forest and may move down below the frost line in winter to avoid freezing. They may be found under logs and other objects on the forest floor, especially after rains.

The yellow spots are beginning to appear on this juvenile spotted salamander, which is several months old.

FOOD AND FEEDING The aquatic larvae eat small invertebrates such as cladocerans, zooplankton, beetles, isopods, ostracods, dragonfly and damselfly larvae, and caddisflies, and sometimes each other. Adults consume a wide variety of invertebrates such as worms, centipedes, millipedes, spiders, and numerous insects, and the occasional small mouse.

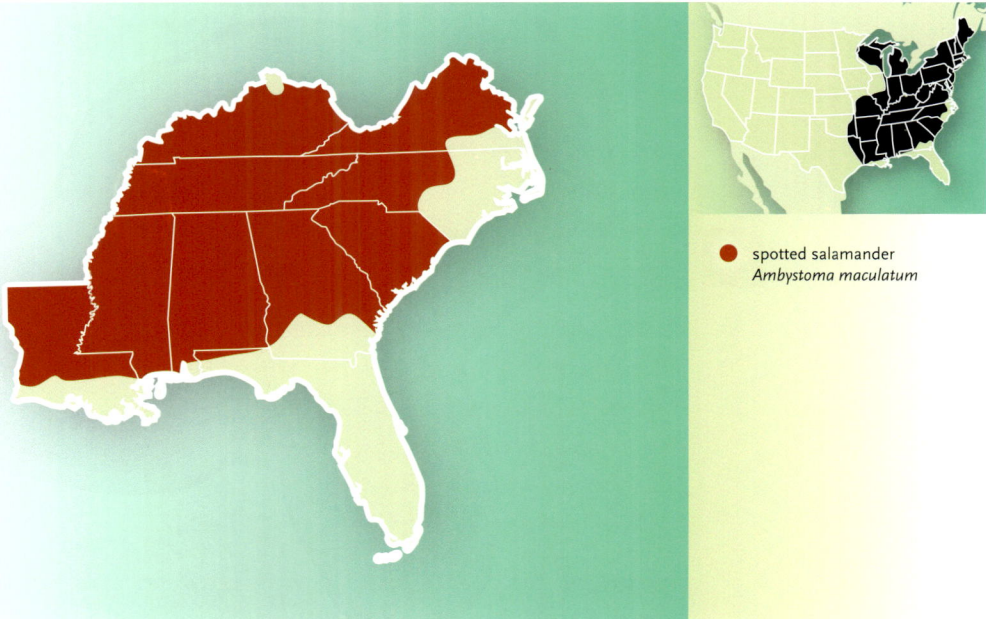

● spotted salamander
  Ambystoma maculatum

Female spotted salamanders lay egg masses that may be either clear or opaque. The difference is based on one gene difference among members of the population.

REPRODUCTION Males arrive first at the breeding pond and deposit several spermatophores (sperm packets) on the substrate. Females arrive shortly afterward and are courted by the males, which use touch and chemical cues to position females over one of their spermatophore packets. The eggs are fertilized as they pass through the female's cloaca while being deposited. The female attaches one to several gelatinous egg masses containing up to 250 eggs to a stem or twig in shallow water. The egg jelly is clear or opaque white. Egg masses that appear green have a symbiotic alga inside individual eggs that may increase the oxygen supply to developing embryos.

PREDATORS AND DEFENSE Various fishes, red-spotted newts, wood frog (*Rana sylvatica*) tadpoles, and aquatic larval and adult invertebrates—including caddisflies, midges, predaceous diving beetles, giant water bugs, water tigers, and dragonfly nymphs—prey on eggs and larvae. Marbled salamander larvae, which are larger, often occur in the same wetlands and will eat spotted salamander larvae.

True albino spotted salamanders with red eyes and no black pigment are rare but may occur anywhere in the species' range.

Wood frog tadpoles, snakes, wading birds, raccoons, and possibly opossums, skunks, and foxes eat juveniles and adults. Egg masses are coated by a thick jelly that may deter some predators; larvae and adults use concealment and noxious skin secretions for defense.

CONSERVATION Spotted salamanders are declining in many areas as a result of deforestation and loss of wetlands, acid precipitation, heavy metal concentrations, environmental contaminants, and fish introduced into the wetlands where they breed. Acid precipitation in the Appalachian Mountains threatens many populations. Effective ways to mitigate elevated acid levels in breeding ponds need to be developed.

TAXONOMY The spotted salamander was first described in 1802. No subspecies are recognized.

Four-toed salamanders are usually reddish above with small, irregular black spots. The constriction at the base of the tail is characteristic of the species.

How do you identify a four-toed salamander?

BODY SHAPE
Stout; constriction around base of tail

NUMBER OF COSTAL GROOVES
13 or 14

BODY COLOR
Reddish brown body with black-and-white belly

PATTERN
Irregular black markings on back; black-and-white belly

BODY SIZE
Adults to 4 inches in total length

Four-toed Salamander   *Hemidactylium scutatum*

DESCRIPTION Four-toed salamanders derive their common name from the four toes on each of the rear feet; most other salamanders have five toes. Adults are usually reddish to rusty brown on the head, body, and tail, and have small, irregular black spots. The sides have gray pigment. The cream to white belly with black spots is unique to this species, and the noticeable constriction at the base of the fragile tail is also a key character.

LARVAE AND JUVENILES The slender larvae are yellowish brown with prominent eyes and a tail fin that extends almost to the head. After metamorphosis, juveniles take on the color and pattern of the adults.

Four-toed salamanders have four toes on each of the rear feet; most other salamanders have five.

The slender, aquatic larvae of four-toed salamanders are yellowish brown with prominent eyes.

SIMILAR SPECIES Dwarf and two-lined salamanders are yellow with a dark stripe on each side and a plain belly. Red and mud salamanders have many or few black spots, respectively, on a red or rusty body, and an unmarked belly.

DISTRIBUTION The highly fragmented range in the Southeast extends from Virginia and eastern Kentucky southward to central Georgia and Alabama. Isolated populations occur in Alabama, Arkansas, Georgia, Kentucky, Louisiana, Mississippi, Tennessee, and northern Florida.

HABITAT Adults and juveniles prefer hardwood forests surrounding swamps, freshwater marshes, isolated woodland pools, and other shallow wetlands without fish. They often remain hidden under rocks, logs, and other surface objects; and under moss on logs. Sphagnum moss is often associated with their breeding ponds.

The cream to white belly with black spots is a signature characteristic of the four-toed salamander.

Sphagnum moss is often associated with the breeding ponds of four-toed salamanders.

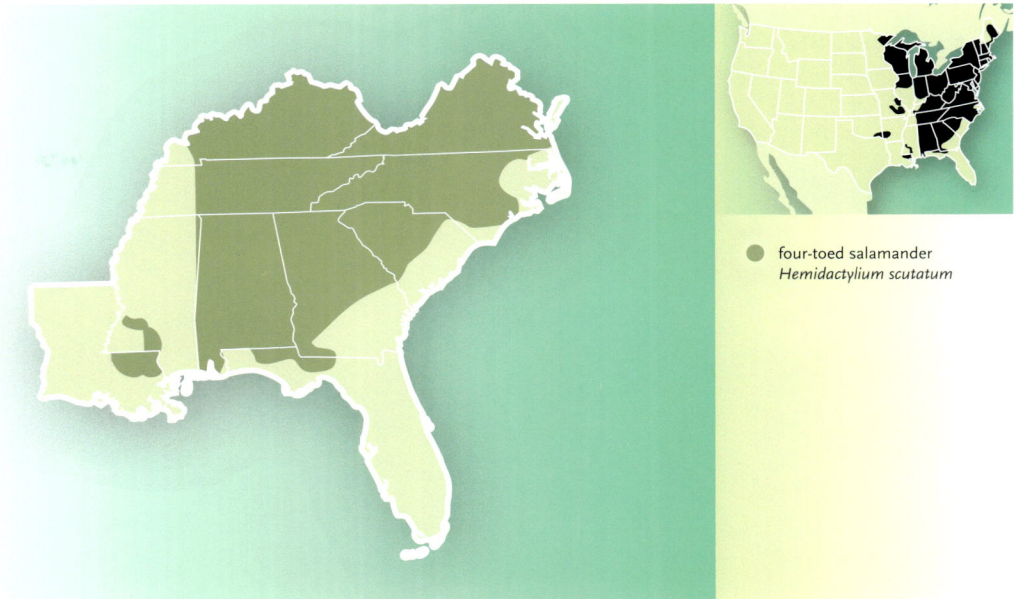

four-toed salamander
Hemidactylium scutatum

BEHAVIOR AND ACTIVITY Adults remain hidden in the forest floor outside the egg-laying season but are occasionally found under logs and other surface objects. Females move to nesting ponds in late winter and return to their terrestrial underground retreats after their eggs hatch. Adults move on rainy nights in spring and summer and are occasionally found on roads.

FOOD AND FEEDING Larvae probably eat small aquatic invertebrates, but little is known about this stage of the life cycle. Juveniles and adults prey on small beetles, moths, spiders, mites, springtails, and worms captured in the forest leaf litter.

Female four-toed salamanders often remain with the eggs until they hatch and drop into the water below the sphagnum moss.

REPRODUCTION Mating occurs on land in fall and early winter, but females do not lay eggs until they migrate to wetlands in late winter and early spring. Each female lays 4–80 eggs in a crude cavity in a clump of moss at the waterline, under loose bark, or in a clump of grasses, attaching single eggs to rootlets and moss strands to form loose clusters. Several females sometimes lay their eggs together in one moss clump, and one or two of the females will remain with the eggs until they hatch and drop into the water below. Reproduction occurs earlier in southern populations than at higher latitudes and elevations. The larval period lasts 4–6 weeks, and metamorphosis occurs in late spring when the larvae are about an inch long.

Juvenile four-toed salamanders take on the color and pattern of the adult after metamorphosis.

Defensive behaviors of the four-toed salamander include coiling the body tightly and breaking the tail completely off at the constriction point. The severed tail distracts the potential predator, allowing the salamander to escape.

PREDATORS AND DEFENSE Larval four-toed salamanders fall victim to predatory insects and the larvae of other salamanders such as marbled, spotted, and Jefferson salamanders, and newts. They do not have noxious skin secretions and are readily eaten by fish, a primary reason why adults avoid breeding sites with fish. Adults are eaten by shrews, skunks, small birds, and ringneck snakes. Defensive behaviors include coiling the body tightly and hiding the head, immobility, emitting noxious secretions from glands in the tail, and breaking the tail completely off at the constriction point. The severed tail distracts the potential predator and allows the rest of the salamander to escape. A new tail is regenerated soon afterward.

CONSERVATION Loss of wetland breeding sites and hardwood forest surrounding them is the main threat to four-toed salamanders. Commercial moss collecting is an underappreciated threat. This species is not protected in any state in the Southeast except by regulations governing their collection.

TAXONOMY The four-toed salamander was first described in 1838. No subspecies are recognized.

Adult eastern newts have distinct red spots on the back bordered by black and a yellow belly with small black spots.

Eastern Newt

Notophthalmus viridescens

DESCRIPTION Adult eastern newts are brown, greenish brown, yellowish brown, olive green, or nearly black; distinct red spots on the back are surrounded by black or reddish lines bordered by black. The belly is yellow with small black spots. A black line runs horizontally through each eye. During the breeding season the male's tail fins increase in height, and black calluses appear on the inside of each hind leg. Females and nonbreeding males have small, almost inconspicuous tail fins and no calluses. Adult newts living aquatically in ponds and other wetlands have smooth, mucous skin; adults that spend a portion of the year in the forest can develop dry, granular skin. The four subspecies differ in color and pattern. Red-spotted newts have red spots on the back, broken-striped newts have partial red lines on the back, and central and peninsula newts lack spots and lines. Peninsula newts are dark olive to nearly black on the back with heavy spotting on the belly.

A black line runs horizontally through each eye.

How do you identify an eastern newt?

BODY SHAPE
Sleek; tail flattened side to side

NUMBER OF COSTAL GROOVES
None

BODY COLOR
Brown to olive green

PATTERN
Red spots or broken reddish lines on the back

BODY SIZE
Metamorphs 2 inches; adults to 4.8 inches in total length

LARVAE AND JUVENILES Larval newts are greenish with a dark stripe that extends from the snout through the eye to the external gills. The body, head, and tail of the terrestrial juveniles, called red efts, are bright yellowish red, red, or orange-red. Terrestrial adults resemble juvenile red efts in shape and skin texture, but their skin color is light to dark brown rather than red.

SIMILAR SPECIES Striped newts are similar in appearance but have two thin, reddish stripes along the back that extend onto the tail.

DISTRIBUTION Eastern newts occur throughout all of the southeastern states.

HABITAT The aquatic adults and larvae live in freshwater temporary and permanent ponds, lakes, ditches, reservoirs, swamps, bogs, canals, and slow-moving streams, especially where there is abundant vegetation. The presence of fish may not pose a problem because newts are toxic. Efts and terrestrial adults live primarily in hardwood forests but may be found in nearby fields and pine forests.

BEHAVIOR AND ACTIVITY Adult eastern newts can be fully aquatic and remain in the larval pond for the remainder of their lives; or they may migrate to the forest and remain there during the nonbreeding season. If the pond

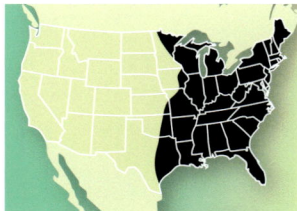

The dark stripe extends from the snout through the eye to the external gills of a larval eastern newt (top).

The body, head, and tail of the terrestrial juveniles (bottom), called red efts, are brightly colored.

eastern newt
Notophthalmus viridescens

red-spotted newt

broken-striped newt

peninsula newt

central newt

Eastern newts occur throughout all of the south-eastern states. An adult and juvenile are shown here.

dries up, the aquatic adults emerge and find moist shelter until the next breeding season or the pond fills again. Mating takes place from late fall through winter in the Southeast. Larvae are light-sensitive and are active in the water column mostly at night; adults are active day and night. Efts are more commonly found walking on the forest floor at night, especially during and after rains, but they are also active during daytime.

FOOD AND FEEDING Larvae eat a wide variety of small aquatic invertebrates, mostly cladocerans, clams, snails, and worms. Terrestrial efts consume adults and larvae of small insects as well as worms, leeches, snails, spiders, and springtails. Aquatic adults eat a wide array of worms, leeches, crustaceans such as grass shrimp, small amphibians and their eggs and larvae, and small fish.

REPRODUCTION Adults perform an elaborate courtship ritual in water. A male grasps a female behind the head with his hind limbs with the goal of enticing her to pick up a sperm packet with her cloaca. Eggs are fertilized internally with sperm from the packet. The female then lays each of her 50–300 eggs singly on a small leaf or stem of an aquatic plant, sometimes over a period of several weeks. Each egg is wrapped in a leaf and glued shut with secretions from the female's cloaca. Eggs hatch into aquatic larvae that remain in the water for 2–5 months. Metamorphosis occurs in August and September, and the juvenile efts remain in the surrounding forest for

The male eastern newt at the top is mating with a female. During the breeding season, the males' tail fins become larger.

up to 7 years before transforming into the adult stage and returning to the pond to breed. In some areas of the Southeast where suitable terrestrial habitat is not available, the larvae skip the eft stage and transform directly into the adult form, possibly retaining their gills as mature adults.

PREDATORS AND DEFENSE Newts secrete toxins from their skin glands at all stages of their life history that make them unpalatable to most predators. Their bright colors warn potential predators of their toxicity. Birds will not eat the brightly colored efts, and fish and most predaceous insects will not eat the adults. Bullfrogs, snapping turtles, painted turtles, garter snakes, and hog-nosed snakes eat an occasional adult but seldom eat the efts, presumably because they are 10 times more toxic than adults. When picked up or otherwise threatened, some newts display a peculiar behavior known as the "unken reflex," in which the newt bends its head backward while simultaneously lifting its tail, forming a circle with its body that prominently displays the brightly colored belly.

When picked up or otherwise threatened, newts will often display a peculiar behavior known as the "unken reflex," in which the newt bends its head backward while simultaneously lifting its tail.

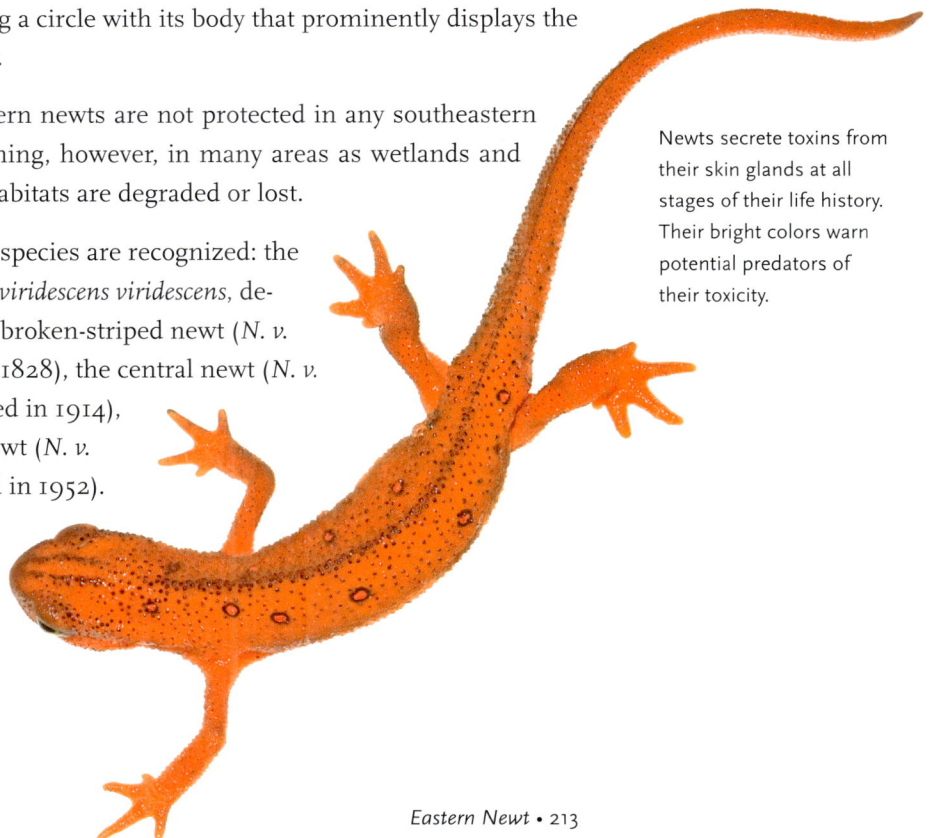

CONSERVATION Eastern newts are not protected in any southeastern state. They are declining, however, in many areas as wetlands and surrounding forest habitats are degraded or lost.

TAXONOMY Four subspecies are recognized: the red-spotted newt (*N. viridescens viridescens*, described in 1820), the broken-striped newt (*N. v. dorsalis*, described in 1828), the central newt (*N. v. louisianensis*, described in 1914), and the peninsula newt (*N. v. piaropicola*, described in 1952).

Newts secrete toxins from their skin glands at all stages of their life history. Their bright colors warn potential predators of their toxicity.

Striped newts usually have red dots on each side of the body and a yellow belly with black specks.

How do you identify a striped newt?

BODY SHAPE
Slender body; fleshy tail

NUMBER OF COSTAL GROOVES
None

BODY COLOR
Tan to dark brown

PATTERN
Irregular series of reddish stripes or small spots on the back and sides

BODY SIZE
Metamorphs 2 inches; adult to 4 inches in total length

Striped Newt

Notophthalmus perstriatus

DESCRIPTION Adult striped newts have rough, tan to dark brown skin. Other characters that may be present include reddish dorsolateral lines thinly bordered by black running the length of the body and extending onto the tail, a row of red dots or streaks on each side of the body, and a faint reddish line along the middle of the back. The belly is yellow with black specks. The back of the cloacal vent of mature males is orange year round. Males in the breeding season have fleshy tail fins, and their inner thighs have black ridges and pads.

Efts of the striped newt are dull in color with reddish stripes on the back and rough skin.

LARVAE AND JUVENILES The aquatic larvae are olive to yellowish green with a black stripe that extends from the snout through the eye to the gills and two irregular, broad dark stripes along each side of the body and tail. Efts are dull, tannish brown with reddish stripes on the back and rough skin.

striped newt
Notophthalmus perstriatus

The aquatic larvae of the striped newt are olive to yellowish green with a black stripe that extends from the snout through the eye to the gills.

SIMILAR SPECIES The broken-striped form of the eastern newt has intermittent red striping on the back, and the stripes are bordered boldly by black. Efts of that species usually have dark spots on a yellow to red body.

DISTRIBUTION Striped newts occur from the Savannah River in southeastern Georgia to north-central Florida, and as far west as the Apalachicola National Forest in Florida and Baker County in southwestern Georgia.

HABITAT This salamander occurs in freshwater ponds, limestone sinkhole ponds, and ephemeral natural ponds that lack fish and are on well-drained sandy soils such as those in shrub and sandhill regions. It prefers such aquatic habitats within pine forests and pine-palmetto stands.

BEHAVIOR AND ACTIVITY Adults live in the upland habitats until rains fill the breeding pond, usually in winter and early spring, then migrate hundreds of yards to the pond to mate and lay eggs. Larvae metamorphose in 3–6 months, depending on rainfall and whether wetlands dry up or

The striped newt has been proposed for protection under the U.S. Endangered Species Act because the ephemeral wetlands in which it breeds are being lost and upland forests are being converted to pine plantations and agriculture.

remain moist. They emerge as efts that spend the next several years in the surrounding forests.

FOOD AND FEEDING Larvae eat a wide variety of aquatic insects and their larvae, as well as other invertebrates. Adults eat worms, snails, spiders, frog eggs, and many types of insects and their larvae.

REPRODUCTION The male courts females to encourage them to pick up one of the spermatophores he has deposited on the pond substrate. Sperm from the packet is used to fertilize eggs as they are deposited singly on aquatic plants. The female wraps each egg in a single leaf that she glues shut with her cloacal secretions. The number of eggs laid by a single female is unknown. The larval period usually lasts 3–6 months but may be several months longer. Some populations remain paedomorphic if wetlands do not dry. The larvae mature into gilled adults, skipping the eft stage.

PREDATORS AND DEFENSE No predators have been documented. Toxic secretions from skin glands probably deter most potential predators.

CONSERVATION This species has been proposed for protection under the U.S. Endangered Species Act because the ephemeral wetlands in which it breeds are being lost and upland forests are being converted to pine plantations and agriculture. Clearing of forest cover around breeding ponds eliminates striped newt populations. Research on restoration of altered habitats and translocation and repatriation into areas where this species once occurred may help enhance the conservation status of this rare species. Currently viable populations and habitats should be protected by modifying silviculture practices.

TAXONOMY The striped newt was first described in 1941. No subspecies are recognized.

FOREST TERRESTRIAL SALAMANDERS

This green salamander from Tennessee has the squared-off toes and lichen-like body pattern characteristic of the species.

How do you identify a green salamander?

BODY SHAPE
Flattened head and body; splayed legs; large feet

NUMBER OF COSTAL GROOVES
14 or 15

BODY COLOR
Black

PATTERN
Profuse green, lichen-like pattern on the head, back, and tail

BODY SIZE
Hatchlings 0.8 inches; adults to 5.5 inches in total length

Green Salamander

Aneides aeneus

DESCRIPTION Adults have a flattened head and body, and the long legs have large feet and squared-off toes. The body is black with abundant green to yellowish green markings on the head, back, and tail that resemble lichens. The belly is pale and unmarked.

LARVAE AND JUVENILES Hatchlings and juveniles are colored and patterned as the adults, but the lichen-like, greenish pattern is more diffuse, and more of the black background is visible.

SIMILAR SPECIES No other southeastern salamander in the range of the green salamander has green markings.

DISTRIBUTION This species is found in the Appalachian Mountains from Pennsylvania southward through southwestern Virginia and eastern Kentucky to northern Alabama and northeastern Mississippi. Isolated populations occur in western North Carolina and South Carolina.

HABITAT Until recently, the green salamander was thought to be restricted to moist crevices in sandstone, granite, and schist rock formations—habitats the flattened body is well adapted to occupy. Recent discoveries indicate that this species may originally have been an arboreal salamander adapted

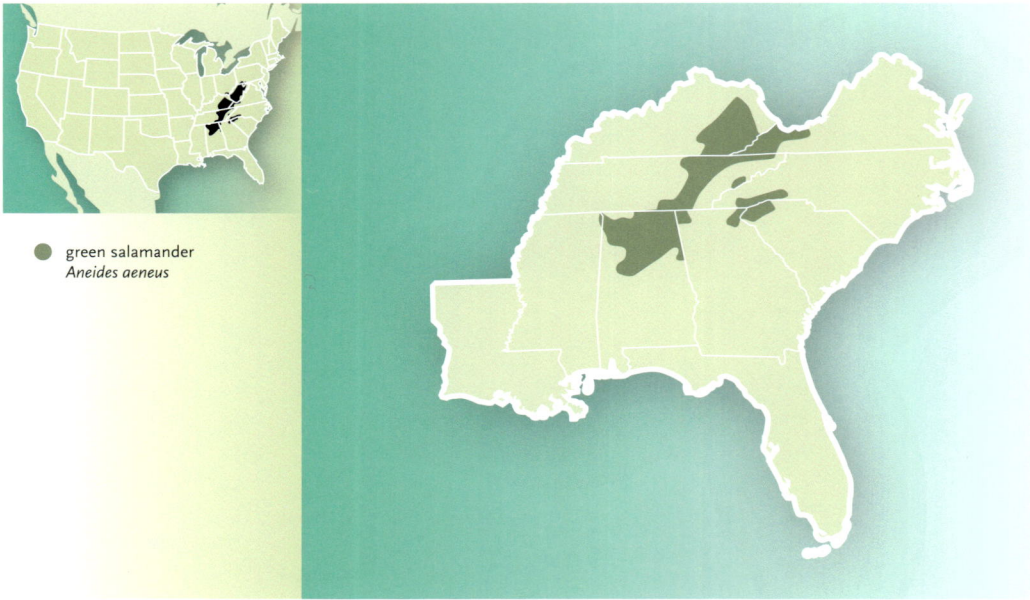

green salamander
Aneides aeneus

to living in tall hardwood trees and under bark; large-scale deforestation by humans may have forced it to move to rock crevices. Individuals are occasionally found under rocks and under bark of logs on the forest floor and in moist, human-made structures such as trail bridges well away from rock formations.

BEHAVIOR AND ACTIVITY Green salamanders are nocturnal. They forage in the open on rock faces and trees during rains in all but the coldest winter months. Males defend their rock crevice territories from other males by biting them, pressing down on their back, and flipping them over. Females are not aggressive toward other green salamanders. Both sexes remain hidden in crevices during the day and retreat deeper as surface moisture levels and temperatures decrease.

FOOD AND FEEDING Adults eat many types of invertebrate prey, including beetles, spiders, flies, crickets, bugs, moths, pseudoscorpions, and mites. Mites appear to be an important prey item for small juveniles.

REPRODUCTION Mating occurs in rock crevices and trees primarily in May and June, with a second peak in the fall. Shortly after mating, females attach their 10–27 eggs to the roof of a rock crevice, under bark on fallen logs, or even high in trees in small clusters by suspending the eggs with adhesive strands of mucus that they produce while lying on their back. Mothers remain with their eggs until they hatch. All of the development occurs in the egg; there is no free-swimming larva. Hatchlings emerge from eggs

Green salamanders remain hidden in rock crevices during the day.

after a 3–5-week incubation period and disperse to other crevices or trees with loose bark.

PREDATORS AND DEFENSE Predators include ringneck snakes, garter snakes, and likely some birds that pull these salamanders out of rock crevices. Green salamanders usually remain immobile when disturbed but may wag their tail from side to side at the mouth of the crevice. The potential predator attacks the tail, which the salamander can break off and later regenerate.

Green salamanders are nocturnal. They forage in the open on rock faces and trees during rains in all but the coldest winter months.

CONSERVATION Population declines have been reported for the Blue Ridge Mountain populations, but this species appears to be stable in other parts of its range. Green salamanders may have been more abundant in the first half of the twentieth century before the introduced chestnut blight killed millions of chestnut trees in the early 1900s. While the salamanders may once have lived under the thick bark of large dead chestnut trees, the logs available today are smaller and have thinner bark. The species is listed as Endangered in Mississippi, Protected in Georgia, and Rare in North Carolina, and is a Species of Special Concern in South Carolina.

TAXONOMY The green salamander was first described in 1881. No subspecies are recognized.

In addition to their tiny body size, pygmy salamanders have distinctive coppery eyelids.

How do you identify a pygmy salamander?

BODY SHAPE
Slender with a short tail

NUMBER OF COSTAL GROOVES
13 or 14

BODY COLOR
Brown to silver

PATTERN
Back has a wide, brownish stripe with a herringbone pattern

BODY SIZE
Adults to 2 inches in total length

Pygmy Salamander   *Desmognathus wrighti*

DESCRIPTION The pygmy salamanders are aptly named because adults do not exceed 2 inches in total length. The broad, reddish brown to tan stripe along the back is embedded within a herringbone pattern of black chevrons that breaks up on the tail. The sides below the stripe are silvery. The belly is flesh-colored, and on animals in the southern part of the range has a pattern of iridescent golden ellipses. The top of the head is rugose, and the tail is rounded in cross section and makes up less than half the total length. The copper-colored eyelids are a distinguishing character. Adult males have a U-shaped mental gland.

LARVAE AND JUVENILES Hatchlings have four or five pairs of light spots on the back and a short tail. Juveniles exhibit adult coloration and patterns. There is no aquatic larval stage.

SIMILAR SPECIES Pygmy salamanders often resemble juveniles of other species of dusky salamanders. The top of the seepage salamander's head is smooth, and the mental gland of males is kidney-shaped. Southern red-backed salamanders lack the eye-to-jaw stripe characteristic of dusky salamanders and do not have the copper-colored eyelids of pygmy salamanders. Ocoee salamanders lack golden ellipses on the belly, and the stripe on the back forms a zigzag pattern.

pygmy salamander
Desmognathus wrighti

DISTRIBUTION This salamander occurs at high elevations from Whitetop and Mount Rogers in southwestern Virginia southward to the Great Smoky Mountains in western North Carolina and eastern Tennessee.

HABITAT Pygmy salamanders can be found in leaf litter and under rocks and decaying logs in and adjacent to seepage areas in high-elevation spruce and fir forests, although they are also commonly found at lower elevations in hard-wood forests in the Great Smoky, Cowee, and Nantahala mountains. They also use moist forest floor habitats well away from springs and seeps. These salamanders aggregate in winter in seepage areas and stream banks, but in warm weather disperse into the surrounding forest.

Pygmy salamanders have no aquatic larval stage, and hatchlings have four or five pairs of light spots on the back.

BEHAVIOR AND ACTIVITY Pygmy salamanders are active in all but the coldest months. Males disperse from winter hibernacula in late spring and forage in terrestrial sites. Females either remain at their winter sites or disperse to stream banks or seepage areas to nest. Adults and juveniles climb trees and other vegetation on wet nights to forage for insects and avoid the larger predatory salamanders on the forest floor. Pygmy salamanders are the most terrestrial of the dusky salamander group.

FOOD AND FEEDING Adults and juveniles eat small invertebrates such as springtails, moths, beetles, flies, spiders, and mites that they catch on rainy nights while perched on tree trunks and herbaceous vegetation on the forest floor.

REPRODUCTION Courtship and mating occur in September and October and in April and May. Shortly after mating, females lay 3–8 eggs under the banks of headwater streams and underground retreats. They attach the eggs to the substrate with a single pedicel formed from the egg capsules and remain with them until hatching. All larval development occurs within the egg. Hatchlings have no external gills.

PREDATORS AND DEFENSE Spring salamanders and ground beetles are known predators. Pygmy salamanders remain immobile when exposed, making them less noticeable to potential predators. These small salamanders avoid open water and stream banks where larger, predatory salamanders occur.

Did you know?

Acid precipitation resulting from human industrial activities has killed spruce and fir trees in the Appalachian Mountains, causing population declines in high-elevation species such as pygmy and seepage salamanders.

CONSERVATION The mountaintop spruce and fir forests preferred by pygmy salamanders are dying as a result of acid precipitation. Salamander populations at elevations above 5,000 feet may be declining in part for that reason and possibly also because warmer temperatures and drier conditions resulting from climate change may exceed their tolerance limits. North Carolina, Tennessee, and Virginia recognize the pygmy salamander in the category of Special Concern as a result of habitat loss and alteration. The isolated populations in the northern end of the range have less genetic variation and more inbreeding than southern populations, suggesting that they are smaller and more vulnerable to extinction, especially if climate change alters high-elevation habitats.

TAXONOMY The pygmy salamander was first described in 1936. Recent molecular research revealed two lineages separated by the French Broad River. Animals from populations north of the river have a slightly longer body and tail, a wider head, and are more robust than salamanders found south of the river. Differences in seasonal temperatures and rainfall between the two regions may be associated with these morphological differences. The population groups north and south of the French Broad River may eventually be designated separate species.

Red Hills salamanders, found only in Alabama, can reach more than 10 inches in total length. They have a dark brown body with no pattern and a thick tail.

Red Hills Salamander   *Phaeognathus hubrichti*

DESCRIPTION These long-bodied salamanders have short legs and a thick, prehensile tail; the total length of adults averages more than 6 inches and can surpass 10 inches. A paler brown area around the snout and head and a paler color on the bottom of the feet are the only contrasts with the uniformly dark brown body. The light line from the eye to the angle of the jaw characteristic of the closely related dusky salamanders is absent.

LARVAE AND JUVENILES The eggs develop directly, with no aquatic larval stage, into terrestrial salamanders that resemble adults in body shape and color.

SIMILAR SPECIES All of the dark-bodied salamanders found in the geographic range of the Red Hills salamander have 16 or fewer costal grooves; the Red Hills salamander has at least 20. The light line behind the eye that is present in all *Desmognathus* species is absent in the Red Hills salamander.

DISTRIBUTION As its common name suggests, this species occurs in the Red Hills physiographic region of southern Alabama. It is found in several counties between the Alabama and Conecuh rivers surrounding Butler County, extending in the southeast to Covington County and in the north to Lowndes and Montgomery counties.

How do you identify a Red Hills salamander?

BODY SHAPE
Elongated body; short legs

NUMBER OF COSTAL GROOVES
20–22

BODY COLOR
Dark brown

PATTERN
Solid body color but paler around mouth

BODY SIZE
Adults to 10.2 inches in total length

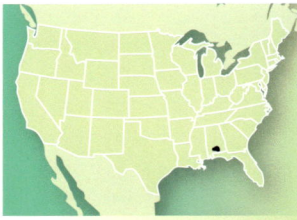

Red Hills salamander
Phaeognathus hubrichti

The Red Hills salamander spends most of its time underground and is most often observed when sticking its head out of a burrow.

HABITAT Hardwood forests with heavy canopies and steep ravines are characteristic habitats, and burrows are more prevalent on the cooler north-facing slopes.

BEHAVIOR AND ACTIVITY The Red Hills salamander spends most of its time underground and is most often observed when sticking its head out of a burrow. More rarely, individuals are seen at night foraging on the ground a short distance from the burrow. Spring and summer are the periods of greatest activity around burrows.

FOOD AND FEEDING These salamanders feed on insects, spiders, snails, and millipedes, ambushing them from the burrow entrance or finding them on the ground nearby at night.

REPRODUCTION Courtship, mating, egg laying, and hatching probably occur underground but have not been observed in nature. The seasonal timing of reproductive events is likewise uncertain. The only information on clutch size is from a captive female that laid 16 eggs. There is no aquatic stage, so all embryonic development must occur within the egg.

PREDATORS AND DEFENSE No predation has been documented, but small snakes and shrews are likely to eat juveniles and adults. Feral pigs and armadillos that dig up burrows are potential predators.

Individual Red Hills salamanders are occasionally seen at night foraging on the ground a short distance from the burrow.

CONSERVATION The Red Hills salamander is listed as Threatened under the U.S. Endangered Species Act. It is the official state amphibian of Alabama and is protected by that state as a Species of Highest Conservation Concern. Its restricted geographic range and specific habitat requirements make the species highly sensitive and susceptible to forestry practices such as clear-cutting and to other activities that degrade hardwood ravine habitats.

The restricted geographic range and specific habitat requirements of the Red Hills salamander make the species highly sensitive and susceptible to forestry practices that degrade hardwood ravine habitats.

TAXONOMY The Red Hills salamander was described as the only member of a new genus in 1961 on the basis of a single specimen discovered in Butler County, Alabama, in 1960. Additional specimens and populations since discovered in the region confirm its uniqueness. No subspecies have been described.

Peaks of Otter salamanders are dark brown to black with brassy flecking and sometimes with small white spots on the sides. The species' distribution is restricted to only two counties in Virginia.

How do you identify a Peaks of Otter salamander?

BODY SHAPE
Slender; long tail

NUMBER OF COSTAL GROOVES
19–22

BODY COLOR
Dark brown to nearly black

PATTERN
Brassy to greenish flecking

BODY SIZE
Adults to 5 inches in total length

Peaks of Otter Salamander *Plethodon hubrichti*

DESCRIPTION Adults are elongated salamanders with a tail approximately the same length as the body. The body and tail are dark brown to nearly black with highly variable amounts of brassy to mosslike greenish flecking. The sides may have small white spots; the belly is plain dark brown to gray.

LARVAE AND JUVENILES Immature individuals, even hatchlings, are colored and patterned as the adults.

SIMILAR SPECIES Red-backed salamanders have a reddish stripe on the back and tail or are nearly black with fine gray speckling on the body and tail; the belly has a salt-and-pepper pattern with white mottling. White-spotted slimy salamanders are black with white spots on the back and sides.

DISTRIBUTION Peaks of Otter salamanders are found only in Bedford and Botetourt counties in the Blue Ridge physiographic province in Virginia. Most of their small range is in the Jefferson National Forest and the Blue Ridge Parkway above 1,500 feet elevation. Some unknown portion of the range extends into adjacent private property.

HABITAT This species prefers mixed hardwood forests with a canopy cover; deep, moist soils; and herbaceous plant cover in spring and summer. A portion of a population will remain in a site from which timber trees were

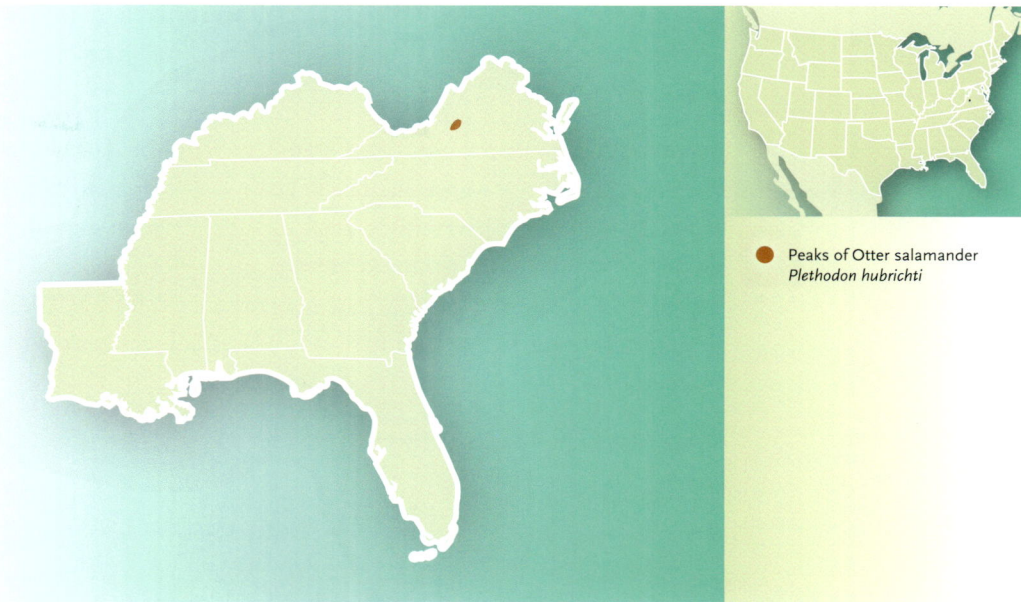

● Peaks of Otter salamander
Plethodon hubrichti

harvested, but the number remaining is highly variable. These salamanders also occur in rhododendron thickets, mixed hardwood and pine forests, and on forested talus slopes.

BEHAVIOR AND ACTIVITY Peaks of Otter salamanders are territorial, but the tiny home range is less than about a square yard. Warm rains will bring them to the surface to forage on the forest floor and on vegetation, including the trunks of trees, up to about 3 feet above the ground. They hide under objects such as logs and rocks during the day and go underground when conditions are dry. The best time to see them active is at night. Only a small portion (about 10 percent) of the population is active on the surface at any given time.

FOOD AND FEEDING Peaks of Otter salamanders eat many types of small invertebrates, including termites, springtails, ants, beetles, spiders, and centipedes.

REPRODUCTION Females lay 5–15 eggs in June and remain with their eggs until hatching occurs in August and September. Otherwise, little has been published on this salamander's courtship, mating, and reproductive biology.

PREDATORS AND DEFENSE Ringnecked snakes, garter snakes, shrews, and wild turkeys and probably other birds that scrape the leaf litter eat these

Peaks of Otter salamanders are territorial, but their tiny home range is only a few square feet.

salamanders. Toxic chemicals in the tail presumably discourage some predators, and remaining immobile when a predator is nearby helps them avoid detection.

CONSERVATION Habitat loss and alteration by removal of hardwood canopy trees appear to threaten populations of Peaks of Otter salamanders, but the population response is variable. The number of salamanders lost depends on how much canopy is removed, how much the forest floor has been disturbed, and whether moist soils persist in the area. A formal federal agency Conservation Agreement and Plan is in place that dictates timber harvest protocols that minimize impacts on this species.

TAXONOMY The Peaks of Otter salamander was first described in 1957. No subspecies are recognized.

Only about 10 percent of the population of Peaks of Otter salamanders is active on the surface at any given time, usually at night.

Shenandoah salamanders occur in two color patterns. One (shown here) has a stripe down the back; the other has no stripe and has a few or no brassy flecks on the back.

How do you identify a Shenandoah salamander?

Shenandoah Salamander *Plethodon shenandoah*

DESCRIPTION Individuals of this small woodland salamander either have a narrow reddish to yellowish stripe down the back on a black background or lack the stripe altogether. If present, the stripe extends from the head onto the tail. Individuals that lack the stripe usually have scattered brassy flecking on the back. The belly of both forms is uniformly black, although some individuals have a few scattered light, irregular markings. The chin is dark with some scattered flecking.

LARVAE AND JUVENILES Immature Shenandoah salamanders are identical with adults and exhibit the same two patterns—striped or unstriped.

BODY SHAPE
Slender

NUMBER OF COSTAL GROOVES
Usually 18

BODY COLOR
Black

PATTERN
Uniform or with a reddish stripe

BODY SIZE
Adults to 4 inches in total length

The belly of the Shenandoah salamander is uniformly black.

The range of the Shenandoah salamander is restricted to rocky slopes on mountaintops in Shenandoah National Park, Virginia.

SIMILAR SPECIES Shenandoah salamanders most closely resemble red-backed salamanders, with which they share habitat in some areas. Red-backed salamanders either have a reddish stripe that is broader than that of the Shenandoah salamander or lack the stripe altogether. The belly of red-backed salamanders has a salt-and-pepper pattern, and the chin is uniformly white to cream. Slimy salamanders in the area are black with small white spots on the back and reach a larger size.

DISTRIBUTION The Shenandoah salamander occurs only on three talus slopes on high mountains (The Pinnacles, Hawksbill, and Stony Man) in the northern part of Shenandoah National Park, Virginia.

HABITAT Shenandoah salamanders occupy shallow to deep soil pockets in talus slopes and the margins of the surrounding hardwood forests. Until recently, they also occupied hemlock stands at high elevations (above 3,000 feet), but infestations of the introduced hemlock woolly adelgid bug have killed the hemlocks and eliminated that habitat.

BEHAVIOR AND ACTIVITY These salamanders are active in all but the coldest months in winter and dry periods in summer, but they remain under surface rocks or logs most of their lives, emerging only on warm, rainy nights to forage. Adults will defend their rocks from intruders, including red-backed salamanders and other Shenandoah salamanders. Most will not move more than a yard or so from their rock their entire lives.

FOOD AND FEEDING Only their mouth size limits the size of prey these salamanders can eat. The diet includes a wide variety of insects, spiders, centipedes, millipedes, and worms. Arboreal insects that alight on the vegetation and rocks on which these salamanders climb on rainy nights are probably eaten as well.

REPRODUCTION Mating occurs on land in the fall and possibly in early spring. Females lay 4–19 eggs every other year under moist rocks deep in the talus and remain with them until they hatch in 4–6 weeks. Few egg clusters have been found.

PREDATORS AND DEFENSE Direct predation has not been observed in nature. Ringneck snakes and shrews are the most likely predators. Defensive behaviors include remaining immobile when uncovered, producing noxious secretions in the tail, and breaking off the tail, which distracts the predator while the rest of the salamander escapes.

CONSERVATION Scientists believe that Shenandoah salamanders are restricted to suboptimal habitat in soil pockets in talus by the more aggressive red-backed salamanders that occupy the rocks in the surrounding moist forest. Only three small, isolated populations are known for this rare species. Droughts and climate warming could cause these populations to shrink and thus become even more severely endangered. This species is listed as Endangered under the U.S. Endangered Species Act and by the state of Virginia. Federal scientists are currently monitoring one or more of the populations.

TAXONOMY The Shenandoah salamander was first described in 1967. No subspecies are recognized.

Did you know?

Unlike frogs, salamanders cannot hear airborne sounds, and most cannot vocalize.

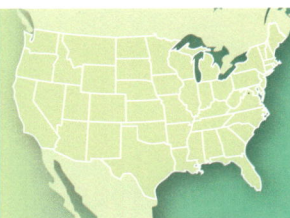

● Shenandoah salamander
Plethodon shenandoah

Yonahlossee salamanders are larger than most plethodontids and have a wide reddish stripe down the back and onto the black tail.

How do you identify a Yonahlossee salamander?

BODY SHAPE
Robust body; stout tail

NUMBER OF COSTAL GROOVES
16

BODY COLOR
Black with a broad, reddish stripe down the back, gray sides, and dark head and belly

PATTERN
Broad, reddish stripe down the back

BODY SIZE
Adults to 8.75 inches in total length

Yonahlossee Salamander   *Plethodon yonahlossee*

DESCRIPTION Yonahlossee salamanders are large and handsome members of the woodland salamander group. The broad red to chestnut stripe down the back and along the base of the tail is characteristic of this species. The stripe is bordered on each side by a narrow, gray to white stripe. The head and limbs are black with light specks; the belly is dark gray with varying amounts of white mottling. Most of the tail is black. Individuals in Rutherford County, North Carolina, have a much reduced reddish stripe that may appear as scattered flecks or blotches.

LARVAE AND JUVENILES Hatchlings and juveniles have four to six pairs of red spots instead of the stripe down the back on a gray body. The spots coalesce with age to form the broad reddish stripe.

SIMILAR SPECIES Other salamanders in the range of the Yonahlossee salamander, such as the red-backed salamander and the mountain dusky salamanders, may have a reddish or yellowish stripe down the back, but the stripe is bordered on each side by dark pigment—never gray or white.

DISTRIBUTION These high-elevation salamanders are found on high ridges (above 1,400 feet) in southern Floyd and Pulaski counties in Virginia, and south through the Blue Ridge Mountains in northeastern Tennessee and western North Carolina almost to the South Carolina line.

Yonahlossee salamander
Plethodon yonahlossee

HABITAT Yonahlossee salamanders prefer virgin and old-growth hardwood forests but may also be found in secondary growth forests with a deep leaf litter layer. They also occur in ravines, on road embankments, on talus slopes, near springs, in pastures, and in rock crevices. When they are not underground in burrows, they seek shelter under logs, limbs, bark, and rocks. Juveniles and adults are found in the same habitats.

BEHAVIOR AND ACTIVITY True seasonal activity is unknown because most of the sites occupied by these salamanders are inaccessible during winter. They probably remain underground until spring temperatures allow movement on the surface without the chance of freezing. They emerge from burrows 1–2 hours after sunset to forage on the forest floor, especially during spring, summer, and fall rains, but remain underground during dry periods in summer. Like other forest-breeding terrestrial salamanders, Yonahlossee salamanders stay within a small area of the forest their entire lives.

Both juvenile and adult Yonahlossee salamanders eat a wide variety of invertebrate prey.

FOOD AND FEEDING Juveniles and adults eat a wide variety of prey, with only their size limiting what they can take. The diet includes snails, spiders, millipedes, centipedes, beetle

Yonahlossee Salamander • 235

adults and larvae, caterpillars, fly larvae and adults, ants, bugs, pseudoscorpions, and worms. Adults occasionally eat other salamanders.

REPRODUCTION Courtship and mating are terrestrial, with the male enticing the female to pick up a spermatophore with her cloaca. Mating likely occurs after emergence from hibernation in April or May, and may also occur in August and September. The female lays about 19–27 eggs in spring and early summer in an underground cavity and probably remains with them until they hatch in August and September.

Young Yonahlossee salamanders have red spots instead of the stripe down the back, but the spots coalesce with age to form the broad reddish stripe seen in adults.

PREDATORS AND DEFENSE Predators have not been identified but are likely to include snakes and birds such as crows and turkeys that forage in the forest litter. Like other woodland salamanders, Yonahlossee salamanders release noxious secretions from glands in the tail that deter some predators. Immobility when first exposed makes them less noticeable to visual predators.

CONSERVATION This mountain species is not protected in any state in which it occurs, although many populations are in national parks and on some Forest Service lands where collecting is prohibited. Logging for timber outside national parks on Forest Service and private lands destroys the canopy and allows sunlight to dry the forest floor, and such habitat alteration is the primary threat.

TAXONOMY This species, described in 1917, was named for Yonahlossee Road on Grandfather Mountain in North Carolina where it was first discovered. The population associated with the Bat Cave and Hickorynut Gorge area in Henderson and Rutherford counties, North Carolina, was described as a separate species in 1962 on the basis of its different color pattern, but genetic evidence indicates that these individuals are variants of the Yonahlossee salamander.

This northern gray-cheeked salamander has a solid gray body and tail just like those of the other three species of gray-cheeked salamanders. Genetic analysis and geographic location are the most reliable means of distinguishing among the four species.

How do you identify a gray-cheeked salamander?

BODY SHAPE
Slender; bulging eyes

NUMBER OF COSTAL GROOVES
16

BODY COLOR
Uniform dark gray

PATTERN
None or rarely with white spots and red flecks; gray patch on each side of the head

BODY SIZE
Adults to 6 inches in total length

Gray-cheeked Salamanders

Blue Ridge Gray-cheeked Salamander   *Plethodon amplus*
South Mountain Gray-cheeked Salamander   *Plethodon meridianus*
Southern Gray-cheeked Salamander   *Plethodon metcalfi*
Northern Gray-cheeked Salamander   *Plethodon montanus*

DESCRIPTION The gray-cheeked salamander group includes four species of woodland salamanders formerly included in the Jordan's red-cheeked salamander (*Plethodon jordani*) complex. All are small (less than 6 inches in total length) with a uniformly dark gray body and gray pigment behind the eye on each side. The Blue Ridge gray-cheeked salamander has a darker belly than the other species in this group. The South Mountain gray-cheeked salamander is larger than the others by about 0.5 inch and has a dark belly and light chin. The southern gray-cheeked salamander has a light gray belly in the northern portion of its range and a dark belly in the southern portion; individuals in some southern populations have white spotting on the sides and brassy flecking on the back. The northern gray-cheeked salamander has a light gray belly and even lighter chin. All four species have grayish cheeks and lack red and white pigment; geographic location is the best character to help identify and differentiate among them.

LARVAE AND JUVENILES Juveniles resemble the adults in their respective ranges.

SIMILAR SPECIES Cumberland Plateau, slimy, and southern Appalachian salamanders all have white to brassy spots on a dark body. Yonahlossee salamanders have a broad reddish stripe down the back and tail. The unstriped phases of eastern and southern red-backed salamanders superficially resemble gray-cheeked salamanders, but both forms have a salt-and-pepper-patterned belly.

Adult South Mountain gray-cheeked salamanders are distinguishable from other gray-cheeked species by the combination of noticeably larger size and a light-colored chin.

DISTRIBUTION Blue Ridge gray-cheeked salamanders are found in the Blue Ridge Mountains in Buncombe, Henderson, and Rutherford counties, North Carolina. South Mountain gray-cheeked salamanders are found in the South Mountains in Burke, Cleveland, and Rutherford counties, North Carolina. Southern gray-cheeked salamanders live in the Blue Ridge Mountains in Haywood and Macon counties, North Carolina; and Oconee County, South Carolina. Northern gray-cheeked salamanders occur in southwestern Virginia from Giles County south through the Valley and Ridge and Blue Ridge physiographic provinces to Haywood County, North Carolina. They occur on Flat Top, Buckhorn, Burkes Garden, Knob, Clinch, and Brumley mountain ridges in the Valley and Ridge province; and on Roan, Bald, Black, Max Patch, and Sandymush mountain ridges in the Blue Ridge.

HABITAT Gray-cheeked salamanders occupy moist high-elevation, full-canopy forests of mixed oaks, maples, buckeyes, ironwood, gum, and tulip poplar; and forests dominated by spruce and fir. They prefer intact substrates and abundant ground cover in the form of rocks and logs.

BEHAVIOR AND ACTIVITY Gray-cheeked salamanders are active on the surface at night and during rainy weather from mid-May to October but work their way underground as freezing temperatures approach in the fall. Populations at lower elevations may become active on warm, wet days in winter, but high-elevation populations remain underground all winter. Adults and juveniles emerge from their extensive burrow systems on warm, rainy nights to forage. Adults defend territories against intruders of their own species.

southern gray-cheeked salamander *Plethodon metcalfi*

both southern gray-cheeked and Blue Ridge gray-cheeked salamanders

Blue Ridge gray-cheeked salamander *Plethodon amplus*

northern gray-cheeked salamander *Plethodon montanus*

South Mountain gray-cheeked salamander *Plethodon meridianus*

both South Mountain gray-cheeked and southern gray-cheeked salamanders

Southern gray-cheeked salamander, *Plethodon metcalfi* (top left); Blue Ridge gray-cheeked salamander, *Plethodon amplus* (top right); northern gray-cheeked salamander, *Plethodon montanus* (bottom left); South Mountain gray-cheeked salamander, *Plethodon meridianus* (bottom right)

The body and tail of southern gray-cheeked salamanders are dark gray. Those in the southern part of the geographic range have a dark belly, and those in more northern areas have a light gray one.

FOOD AND FEEDING Foraging occurs at night on the ground or while the salamander is climbing into herbaceous vegetation and tree trunks as high as 6 feet above the ground. Known prey includes ants, moth and beetle larvae, moths, beetles, flies and their larvae, wasps, stinkbugs, weevils, crickets, leafhoppers, scorpionflies, gnats, springtails, mites, spiders, millipedes, centipedes, and worms.

REPRODUCTION Courtship and mating take place on land, with each male using visual and tactile behaviors to entice a female to pick up a sperm cap from one of his spermatophores. After fertilization, females lay fewer than 10 eggs, probably in underground cavities although no natural nests have been found. Like other woodland salamanders, females likely remain with their eggs until hatching, which probably occurs in late summer to early fall. The hatchlings remain underground for 10–12 months before emerging on the surface the following May or June.

PREDATORS AND DEFENSE Little is known about predators, but they likely include small snakes, shrews, ground-foraging birds, and other salamanders such as the spring salamander. All four gray-cheeked salamander species have glands in the tail that secrete mucus that may be noxious and may deter some small predators. Immobility following exposure or disturbance may cause a visual predator to overlook the salamander.

CONSERVATION These woodland salamanders are threatened with loss of forest habitat from timber operations, pollution from acid precipitation that kills spruce and fir trees at high elevations, and conversion of forested areas into housing and urban zones. None of these species is listed as Endangered or Threatened.

TAXONOMY These four species were considered part of the Jordan's red-cheeked salamander complex until 2000, when the results of genetic analyses revealed each to be a distinct species.

Red-backed Salamanders

Eastern Red-backed Salamander   *Plethodon cinereus*

Big Levels Salamander   *Plethodon sherando*

DESCRIPTION Red-backed salamanders occur in two color phases in the southeastern portion of their range: one has a broad, straight-edged, or-ange, yellowish, or reddish stripe on the back and tail (red-backed phase); the other lacks the stripe (lead-backed phase). Both color phases can occur in the same population, although proportions differ geographically. Body color for both is dark gray to black with small white flecks or spots on the head, sides of the body, and tail; and both have a black belly with tiny white or yellowish flecks forming a salt-and-pepper pattern. A uniformly reddish phase of this species with no stripe occurs in the Northeast and rarely in the Southeast. Big Levels salamanders look like red-backed salamanders but have a wider head, longer limbs, and fewer costal grooves (1–5) between the limbs when they are folded back alongside the body; red-backed sala-manders have 5.5–9 costal grooves between the front and hind limbs.

LARVAE AND JUVENILES Hatchlings and juveniles are identical with the adults of their species but have a shorter tail.

SIMILAR SPECIES Southern ravine and Valley and Ridge salamanders are similar in body color to the lead-backed phase of the eastern red-backed

How do you identify a red-backed salamander?

BODY SHAPE
Slender; long tail

NUMBER OF COSTAL GROOVES
18–20, usually 19

BODY COLOR
Dark gray to black

PATTERN
Reddish stripe on back and tail, or no stripe

BODY SIZE
Adults to 5 inches in total length

salamander, but their bellies are uniformly black. The orange to reddish stripe on southern red-backed salamanders usually has wavy or saw-tooth edges and extends well onto the tail, but this species is not known to co-occur with the eastern red-backed salamander.

DISTRIBUTION Eastern red-backed salamanders are widely distributed in northeastern North America from Minnesota to Maine and south to Kentucky. Southeastern populations occur throughout most of Virginia, much of North Carolina, and the extreme northeastern counties of Tennessee. Big Levels salamanders are limited to a small area in the Blue Ridge Mountains in Augusta and Nelson counties, Virginia.

HABITAT These salamanders prefer deciduous mixed hardwood forests, mixed conifer-hardwood forests, cove forests, and bottomland forests with deep, moist soil–leaf litter layers, where they can reach densities of more than two salamanders per square meter. They occur in lower densities in a wide range of other habitats, including mixed pine and hardwoods, edges of hardwood forests and fields, urban forest patches, and woodlands in agricultural areas.

BEHAVIOR AND ACTIVITY Eastern red-backed salamanders are active in all but the coldest months, although individuals can remain inactive for prolonged periods. Surface activity is greatest during warm rains, when they move about on the forest floor and climb herbaceous vegetation to forage for prey; but only a small portion of the population (2–32 percent) is active on

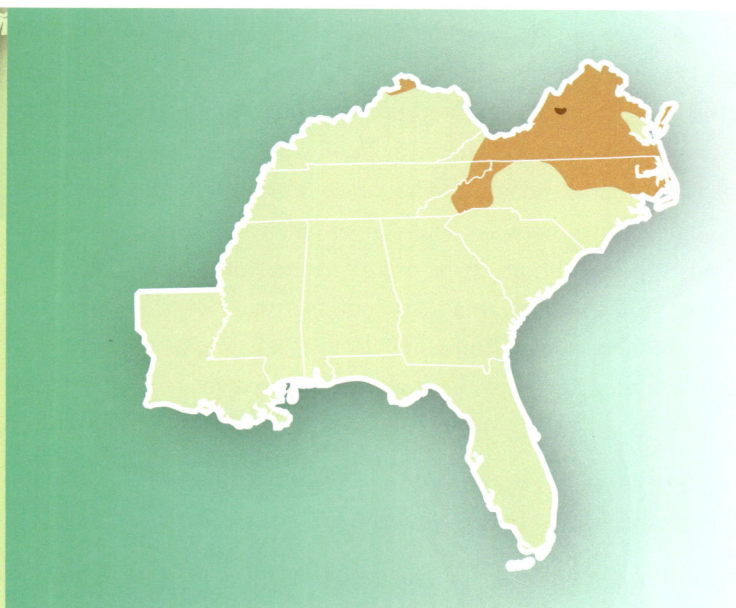

● eastern red-backed salamander
Plethodon cinereus

● Big Levels salamander
Plethodon sherando

Big Levels salamanders (below), which are found only in two counties in the Blue Ridge Mountains in Virginia, look like red-backed salamanders (left) but have a wider head and longer legs.

the surface at any given time. They hide under leaf litter, logs, rocks, and other surface objects when the forest floor is moist but work their way down deep into the leaf litter and soil as the surface dries. These salamanders are not active on the surface in hot and dry weather. Males and females defend their small territories, usually defined by a rock or log, from other red-backed salamanders and can recognize other individuals. Most never move more than a few yards away from their birth site.

FOOD AND FEEDING These salamanders forage on warm, rainy nights on the forest floor or climb herbaceous vegetation and tree trunks where they sit and wait for prey. The diet consists of any prey they can catch and swallow whole, including ants, mites, termites, earthworms, centipedes, snails, slugs, beetles, moths and their larvae, flies, springtails, and midges. Adults sometimes cannibalize eggs and juveniles. Eastern red-backed salamanders in one population were estimated to eat more than 1.5 million prey items per hectare per year. Their extreme abundance in intact forests ensures a huge role in the ecosystem's energy transfer chain.

REPRODUCTION Mating takes place from about November to March. Males court females on land, enticing them to pick up spermatophores. Each female lays 1–14 eggs in a moist cavity in leaf litter, under logs and rocks, inside logs, or in the soil. All development occurs within the egg capsule, and the gills are lost just before hatching in August or September. Female parents usually remain with their eggs until hatching to protect them from small predators.

PREDATORS AND DEFENSE Predators include ringneck snakes, garter snakes, copperheads, shrews, voles, chipmunks, raccoons, foxes, turkeys and other ground-foraging birds, beetles, and praying mantises. Noxious, sticky secretions produced by glands in the tail clog predators' mouths. Red-backed salamanders will also remain motionless to avoid detection, flee, assume a coiled position to protect their head, and break off their tail. The twitching of the broken-off tail distracts the predator while the salamander escapes.

CONSERVATION Eastern red-backed salamanders and Big Levels salamanders are not protected in any state in which they occur because they are widespread and abundant. Local populations are threatened, however, when timber harvesting and other practices eliminate the forest canopy and leaf litter, and fragment their habitat into smaller and smaller patches. Because they do not migrate, these salamanders are affected by any activities in their area. Urbanization can cause complete loss of habitat and extirpation of populations.

TAXONOMY The eastern red-backed salamander was first described in 1818. The Big Levels salamander was described in 2004 largely on the basis of genetic differences from the eastern red-backed salamander. No subspecies are recognized.

The female red-backed salamander lays as many as 14 eggs in a moist cavity and usually remains with the eggs until they hatch.

White-spotted
slimy salamander

Slimy Salamanders

Northern Slimy Salamander	*Plethodon glutinosus*
Chattahoochee Slimy Salamander	*Plethodon chattahoochee*
Atlantic Coast Slimy Salamander	*Plethodon chlorobryonis*
White-spotted Slimy Salamander	*Plethodon cylindraceus*
Southeastern Slimy Salamander	*Plethodon grobmani*
Louisiana Slimy Salamander	*Plethodon kisatchie*
Mississippi Slimy Salamander	*Plethodon mississippi*
Ocmulgee Slimy Salamander	*Plethodon ocmulgee*
Savannah Slimy Salamander	*Plethodon savannah*
South Carolina Slimy Salamander	*Plethodon variolatus*

DESCRIPTION The slimy salamander complex includes 10 species native to the Southeast that were formerly all considered to be *Plethodon glutinosus*. All have similar ecology and behavior. Rely on geographic location for species identification.

All slimy salamanders are black to bluish black. The largest species reach a total length up to 8 inches. Species from the Coastal Plain are at least half an inch shorter than those from the interior. Most have white spots

How do you identify a slimy salamander?

BODY SHAPE
Robust; tail about the same length as the body

NUMBER OF COSTAL GROOVES
Average 16

BODY COLOR
Black to bluish black

PATTERN
White to brassy spots on the back and tail; spots more abundant on the sides

BODY SIZE
Adults to 8 inches in total length

or brassy flecks on the back and tail and a uniformly dark belly. Northern slimy salamanders have large brass-colored spots on the back and tail and moderately abundant white and yellow spotting on the sides of the body. Chattahoochee slimy salamanders have a light chin, little to no spotting on the back and tail, and abundant white to yellow spots on the sides. Atlantic Coast slimy salamanders have small brassy and white spots on the back and tail and abundant white to yellow spotting on the sides. White-spotted slimy salamanders have a light chin, large white spots on the back and tail, and moderately abundant white spots on the sides. Southeastern and Mississippi slimy salamanders have large brassy spots on the back and tail and abundant white to yellow spotting on the sides. Louisiana slimy salamanders have very large brassy to white spots on the back and numerous yellow spots on the sides and legs. Ocmulgee slimy salamanders have few small brassy spots on the back and tail and a moderate amount of white spotting on the sides. Savannah slimy salamanders have very little brassy

Slimy salamanders were once considered a single species but are now divided into 10 different species: (left, top to bottom) Chattahoochee slimy, *Plethodon chattahoochee;* Atlantic Coast slimy, *Plethodon chlorobryonis;* white-spotted slimy, *Plethodon cylindraceus;* southeastern slimy, *Plethodon grobmani;* South Carolina slimy, *Plethodon variolatus;* Mississippi slimy, *Plethodon mississippi;* Ocmulgee slimy, *Plethodon ocmulgee;* (below, top left) Savannah slimy, *Plethodon savannah;* (below, top right) Louisiana slimy, *Plethodon kisatchie;* (bottom) northern slimy, *Plethodon glutinosus.*

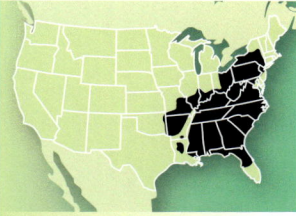

Mississippi slimy salamander
Plethodon mississippi

both Mississippi and
northern slimy salamanders

northern slimy salamander
Plethodon glutinosus

both northern slimy and
white-spotted slimy salamanders

white-spotted slimy salamander
Plethodon cylindraceus

Chattahoochee slimy salamander
Plethodon chattahoochee

South Carolina slimy salamander
Plethodon variolatus

Atlantic Coast slimy salamander
Plethodon chlorobryonis

Savannah slimy salamander
Plethodon savannah

Ocmulgee slimy salamander
Plethodon ocmulgee

southeastern slimy salamander
Plethodon grobmani

Louisiana slimy salamander
Plethodon kisatchie

Although most slimy salamanders move no more than 30 feet from their hatching site, adults defend their small territories from other slimy salamanders and other woodland salamander species.

pigment in the white spots on the back and tail but have abundant white to yellow spotting on the sides. South Carolina slimy salamanders have few brassy spots on the back and tail and a moderate amount of white spotting on the sides.

LARVAE AND JUVENILES Hatchlings and juveniles of each species resemble their parents in color and pattern.

SIMILAR SPECIES Cumberland Plateau salamanders have a light chin, abundant brassy spots on the back, and abundant white to yellow pigment on the sides. Southern Appalachian salamanders resemble slimy salamanders but usually have small red spots on the legs. Tellico salamanders have abundant brassy flecking and spots on the back, but they occur with other slimy salamanders only in Polk County, Tennessee. The lead-backed phase of the eastern red-backed salamander has tiny flecking and a salt-and-pepper belly. Cow Knob salamanders lack brassy or silvery flecking and have small yellow spots on the belly. Some Wehrle's salamanders superficially resemble northern slimy salamanders but have fewer and smaller spots on the back and 17 costal grooves rather than the 16 usually found in slimy salamanders.

DISTRIBUTION Northern slimy salamanders are widespread from New York to Illinois south through southwestern Virginia, most of Kentucky, all but far western Tennessee, and in northern Alabama and Georgia. Chattahoochee slimy salamanders occur in the northern counties of Georgia that lie in the Blue Ridge physiographic province and in Cherokee County, North Carolina. Atlantic Coast slimy salamanders are found in the Coastal Plain physiographic province from southeastern Virginia through most of South Carolina into northeastern Georgia. White-spotted slimy salamanders occur in the Piedmont and Blue Ridge physiographic provinces of Virginia and North Carolina to the French Broad River and south into northern South Carolina. Southeastern slimy salamanders occur in southern Alabama and Georgia south into the central Florida peninsula. Louisiana slimy salamanders are found in central Louisiana north into southern Arkansas. Mississippi slimy salamanders are found from western Kentucky south through western Tennessee into Mississippi and the western half of Alabama. Ocmulgee slimy salamanders are known from central and eastern Georgia. Savannah slimy salamanders are known only from eight counties in east-central Georgia. South Carolina slimy salamanders occur in the southern Coastal Plain of South Carolina and in several counties along the Savannah River in eastern Georgia.

HABITAT Slimy salamanders occur in a wide variety of habitats throughout their range, including mature hardwood forests, ravines, cove forests, old pine stands, and river floodplains. They use numerous surface objects for

All slimy salamanders are black to bluish black and most have white spots or brassy flecks on the back.

Slimy Salamanders • 249

shelter on the forest floor, including rocks, logs, slabs of tree trunk or bark, and leaf litter. They are often associated with decomposing tree stumps. Like other woodland salamanders, they remain in a small area of forest their entire lives.

BEHAVIOR AND ACTIVITY Slimy salamanders in the mountains and in Kentucky, Tennessee, and Virginia are active in all warm months of the year—generally from March through October—if rainfall has been sufficient. Those in the Coastal Plain and in the Deep South are more active during winter months than in the hot, dry summer months. They move underground and into burrows and root tunnels as the surface dries, and sometimes form aggregations under large logs. Populations near caves may move inside when the forest dries out and move back out again during rains. They forage on the surface when it is wet or during warm rains. Most move no more than 10 yards from their hatching site. The maximum distance recorded for movement by a juvenile slimy salamander is 92 yards. Adults defend their small territories from other slimy salamanders and other woodland salamander species.

FOOD AND FEEDING Adults and juveniles emerge from their retreats on wet nights to forage on the forest floor. Appalachian populations are known to feed only in the first few hours after sunset. All slimy salamanders are generalist predators that eat a wide array of prey types. Insect groups include beetles, leafhoppers, aphids, bugs, moths, flies, and ants. Other invertebrate groups include springtails, daddy longlegs, spiders, mites, pseudoscorpions, millipedes, centipedes, worms, snails, slugs, and isopods. They also eat other species of salamanders as well as other individuals of their own species.

REPRODUCTION Male slimy salamanders court females on land using chemical, tactile, and visual behaviors to entice them to pick up a sperm cap from a spermatophore. Following internal fertilization, females lay 10–33 eggs in clusters that they suspend from the ceiling of a small cavity in the ground, under rocks and logs, and in crevices in caves and mine shafts. Females remain with their eggs until they hatch in August through December following a 2–3 month embryonic development period. The timing depends on latitude, elevation, and proximity to the coast.

PREDATORS AND DEFENSE Known predators include copperheads, garter snakes, cave salamanders, and shrews; ground-foraging birds such as turkeys may eat them. The noxious, sticky mucus produced by glands in the tail deters some predators because it tastes bad and glues their mouth shut.

CONSERVATION Hardwood forest habitats used by all slimy salamanders in the group have been subjected to timbering and clearing for agriculture, roads, and urban areas. Conversion of hardwood forests to pine monocultures also eliminates habitat. Other threats include alteration of the forest microenvironment by acid precipitation, climate change, and introduced earthworms.

TAXONOMY The northern slimy salamander, described in 1818, included all of the other slimy salamander species until molecular analyses determined that the complex comprises a number of forms that should be considered full species. The Chattahoochee slimy salamander, Louisiana slimy salamander, Mississippi slimy salamander, Ocmulgee slimy salamander, and Savannah slimy salamander were recognized as species in 1989. The Atlantic Coast slimy salamander (described in 1951), the white-spotted slimy salamander (described in 1825), and the South Carolina slimy salamander (described in 1818) were resurrected in 1989 from what had been considered incorrect taxonomy.

Hardwood forest habitats used by all slimy salamanders in the group have been subjected to timbering and clearing for agriculture, roads, and urban areas.

An individual Valley and Ridge salamander can be distinguished from a Shenandoah Mountain salamander only by genetic analysis or the geographic location of the population from which it came.

BODY SHAPE
Slender; long tail

NUMBER OF COSTAL GROOVES
20 or 21

BODY COLOR
Brownish black to black

PATTERN
Numerous small white spots or brassy flecks, often concentrated on the sides

BODY SIZE
Adults to 5.4 inches in total length

Ridge Salamanders

Valley and Ridge Salamander   *Plethodon hoffmani*

Shenandoah Mountain Salamander   *Plethodon virginia*

DESCRIPTION The two species can usually be distinguished only genetically or by geographic location. Both are elongated salamanders with short legs and a tail that is longer than the body. Adults are brownish black to black with numerous small, scattered white spots and brassy flecks that are usually more concentrated along the sides. Some red pigment may be present on the back, but most individuals have no hint of a pattern; the exception is some Shenandoah Mountain salamanders on Reddish Knob and Shenandoah Mountain along the Virginia–West Virginia border that have narrow reddish stripes. The throat is white and the belly is black with some white mottling.

LARVAE AND JUVENILES Hatchlings have not been described, but juveniles are identical with adults in color and pattern.

SIMILAR SPECIES The lead-backed phase of the eastern red-backed salamander closely resembles both of these species but has a salt-and-pepper belly. Ravine salamanders are even more elongated and have a black belly and light flecking on the throat.

DISTRIBUTION Valley and Ridge salamanders range from central Pennsylvania south through the Valley and Ridge physiographic province through eastern West Virginia to the New River in Virginia. Shenandoah Mountain salamanders are found in western Rockingham County, Virginia, and northward through several counties in West Virginia that border Virginia to the Potomac River in Maryland.

HABITAT Valley and Ridge and Shenandoah Mountain salamanders occupy deciduous hardwood forests on mountain slopes and ravines where flat rocks are abundant. Individuals can be found under moist logs and rocks and in leaf litter, but most of the population remains underground within the interstices of the rock-soil zone. The substrate in their ranges tends to be dry and well drained. Moist pockets of leaf litter and herbaceous vegetation support greater densities than dry areas.

Shenandoah Mountain salamanders live in hardwood forests in rocky, mountainous areas, on slopes and in ravines where flat rocks are abundant. They are most active on the surface during moist periods from April through September.

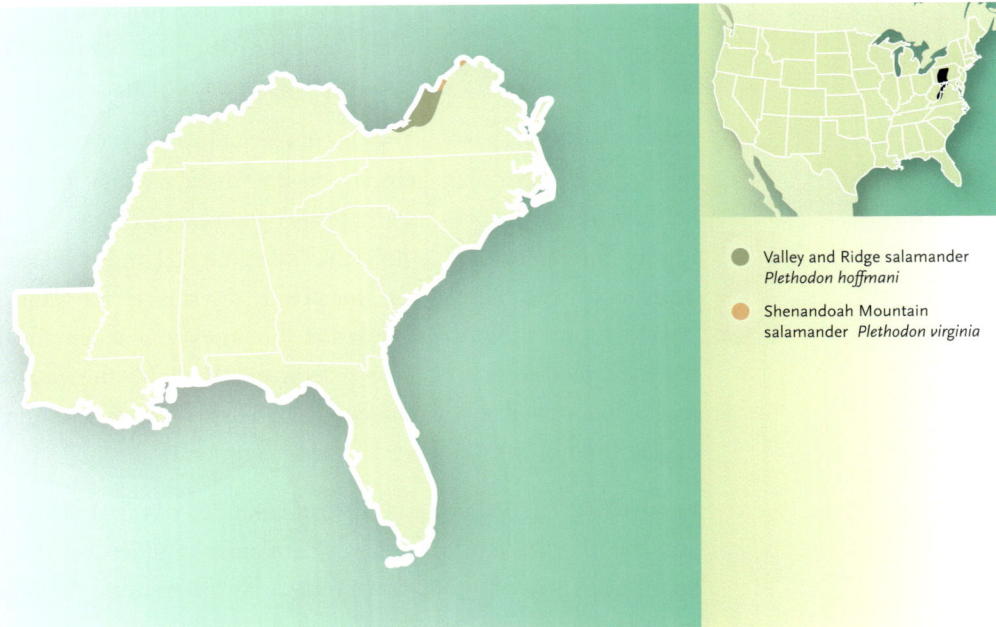

BEHAVIOR AND ACTIVITY These salamanders are not active on the surface in cold winter months, although they may move about below ground. Surface activity occurs at night during warm rains from April through September, when individuals forage in and under leaf litter and climb herbaceous vegetation to

● Valley and Ridge salamander
Plethodon hoffmani

● Shenandoah Mountain
salamander *Plethodon virginia*

catch prey. They are not active in warm summer months when the ground is dry. Like other small woodland salamanders, they spend most of their lives within a very small area of the forest floor.

FOOD AND FEEDING Valley and Ridge and Shenandoah Mountain salamanders consume a wide range of invertebrates including snails, worms, spiders, millipedes, centipedes, collembolans, beetles, flies, ants, and insect larvae. Larger salamanders eat both small and large prey.

REPRODUCTION Mating and courtship take place on land in spring and fall. Males presumably deposit spermatophores and entice females to pick them up, although that behavior has not yet been described for these two species. Females lay three to eight eggs in underground cavities in May and June and likely remain with them until they hatch. Hatching occurs in August and September, but the young do not emerge from underground retreats until the following spring.

The belly of ridge salamanders is black with some white mottling.

PREDATORS AND DEFENSE Natural predators have not been observed, but small snakes, shrews, and ground-foraging birds such as turkeys probably eat these salamanders. Both species produce noxious secretions from glands in the tail that deter small predators. They often remain immobile when first uncovered, although some flee quickly to underground retreats.

CONSERVATION These salamanders are not protected in any state in which they occur. Timber management activities in the national forests and private lands in the region alter their habitats by removing canopy trees and exposing the forest floor to sunlight that dries the soil and leaf litter. This may be especially problematic on the already dry soils of the Valley and Ridge physiographic province. Populations are known to decline in areas where the hardwood forests have been removed. Probably the greatest threat to forest floor–dwelling salamanders is the disturbance to the first few inches of soil and humus that occurs during logging.

TAXONOMY The Valley and Ridge salamander was first described in 1971, and the Shenandoah Mountain salamander was described as a separate species in 1999 on the basis of molecular differences. No subspecies are recognized.

The Cumberland Plateau salamander is superficially similar to the northern slimy salamander, from which it can be differentiated by genetic analysis and geographic location. Its head is also relatively broader and slightly flattened.

How do you identify a Cumberland Plateau salamander?

BODY SHAPE
Slender; bulging eyes

NUMBER OF COSTAL GROOVES
16

BODY COLOR
Black

PATTERN
Scattered white spots on the body and tail that are more concentrated on the sides; light chin and throat

BODY SIZE
Adults to 6.5 inches in total length

Cumberland Plateau Salamander

Plethodon kentucki

DESCRIPTION Cumberland Plateau salamanders are relatively large and have big, bulging eyes. They are entirely black with small white spots scattered on the head, back, and tail and a greater concentration of larger, irregular white spots on the sides. The belly is uniformly slate gray; the chin and throat are light gray. Males have a very large mental (chin) gland.

LARVAE AND JUVENILES Hatchlings and juveniles closely resemble the adults but have fewer white spots.

SIMILAR SPECIES The northern slimy salamander is very similar in appearance, and its range completely overlaps that of the Cumberland Plateau salamander. Slimy salamanders are black with an abundance of silvery white to metallic gold spots on the head, body, and tail; larger spots on the sides; and a grayish black belly, throat, and chin. Adult male slimy salamanders have a smaller chin gland.

DISTRIBUTION Cumberland Plateau salamanders occur in southern West Virginia, the eastern third of Kentucky, southwestern Virginia, and several counties in Tennessee along the Tennessee-Kentucky state line.

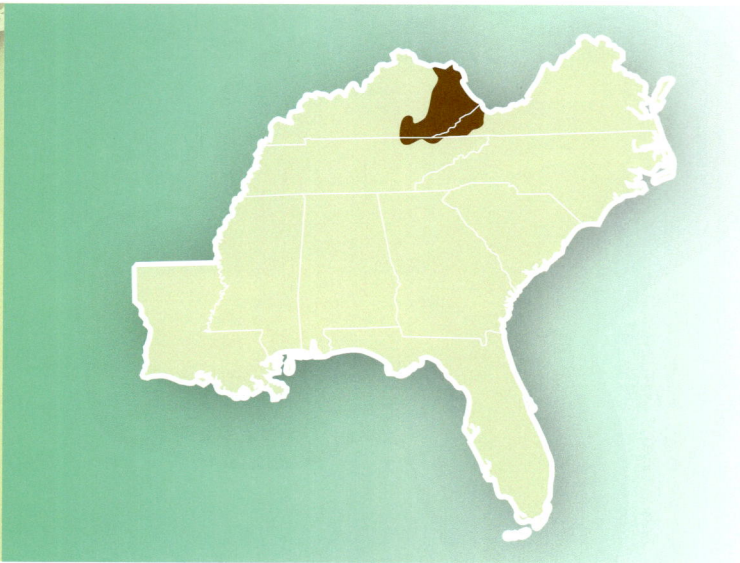

● Cumberland Plateau
salamander
Plethodon kentucki

The Cumberland Plateau
salamander has a light
gray chin and throat.

HABITAT These woodland salamanders prefer mature hardwood forests. Habitats include mixed hardwood forests, moist ravines, hillsides, shale banks, and cave entrances. Individuals use rocks and logs on the forest floor for shelter but often seek hiding places in rock crevices.

BEHAVIOR AND ACTIVITY Adults and juveniles are active on the surface on warm, moist days from about April through October, although they go underground during hot, dry weather. Rain brings them out to forage. They remain in a small area their entire lives; known movement distances are only 1–2 yards.

FOOD AND FEEDING All size classes forage on the forest floor and on low vegetation, eating anything they can subdue and swallow. Prey items include forest floor dwellers such as ants, springtails, beetles, flies, mites, spiders, pseudoscorpions, and snails.

REPRODUCTION Male Cumberland Plateau salamanders, like other *Plethodon* species, court and mate on land in summer and fall. The male uses visual and tactile stimulation to entice a female to pick up the sperm cap from the spermatophore he has deposited on leaf litter. Once fertilization has occurred internally, females deposit 9–12 eggs in clusters in underground cavities in summer and remain with them until they hatch in September and October. Hatchlings remain with their mother for several weeks until their yolk reserves are used up, then move into underground burrows and crevices on their own. They emerge on the forest floor the following June or July when they are 8–9 months old.

Cumberland Plateau salamanders often seek hiding places in rock crevices.

PREDATORS AND DEFENSE Predators of Cumberland Plateau salamanders have not been reported, but snakes and ground-foraging birds undoubtedly eat them. Like slimy salamanders, they produce noxious secretions from glands in the skin that deter small predators.

CONSERVATION This species is not protected in any state in which it occurs. Loss and alteration of mature hardwood forests from timbering operations and conversion to agriculture and urbanization cause populations to decline and disappear.

Adult and juvenile Cumberland Plateau salamanders are active on the surface on warm, moist days from about April through October, although they go underground during hot, dry weather.

TAXONOMY The Cumberland Plateau salamander was first described in 1951 as a full species, but from the mid-1950s until the mid-1980s was thought to be a slimy salamander. Genetic analysis in 1983 showed that the Cumberland Plateau salamander differs from slimy salamanders enough to warrant full species recognition.

Southern red-backed salamanders are small and occur in two color patterns; this is the striped form. All individuals have brown sides, usually with some white markings.

How do you identify a southern red-backed salamander?

BODY SHAPE
Slender; tail round and less than half the total length

NUMBER OF COSTAL GROOVES
18–22

BODY COLOR
Brown with a small amount of reddish and white pigment

PATTERN
Reddish stripe along the middle of the back and tail, or solid gray with no stripe

BODY SIZE
Adults to 4.2 inches in total length

Southern Red-backed Salamander

Plethodon serratus

DESCRIPTION Southern red-backed salamanders occur in both striped and unstriped forms. The striped version has a reddish to orangish, or rarely gray, stripe down the middle of the back and tail. The edges of the stripe are distinctly serrated, and each small projection alternates with a costal groove. The color of the stripe often fades onto the sides, especially along the front of the body. The back and tail of the unstriped form, which is found primarily in Alabama and Georgia, are brownish; the back, sides, and belly have scattered red spots. Both forms have brown sides with varying amounts of white. The tail is round and shorter than the body.

LARVAE AND JUVENILES Juveniles are patterned and colored as the adults.

SIMILAR SPECIES Eastern red-backed salamanders are similar in color and pattern but have a salt-and-pepper belly pattern, and the ranges of the two species do not overlap. The reddish stripe on southern zigzag salamanders is distinctly lobed along the edges rather than serrate, especially on the front half of the body. The orange to reddish stripe on Webster's salamanders has wavy edges, at least on the front half of the body. The latter two species also have 18 costal grooves.

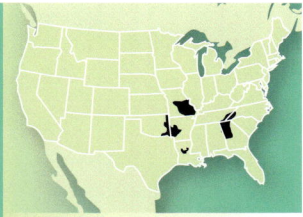

southern red-backed
salamander
Plethodon serratus

Some southern red-backed
salamanders without the
reddish stripe on the back
(lead-backed phase) also
lack the brown color on
each side.

DISTRIBUTION Southern red-backed salamanders occur in four widely
separated areas in western Arkansas, Missouri, central Louisiana, and
southwestern North Carolina to central Georgia.

HABITAT These salamanders occupy a wide variety of habitat types, most with
one or more species of hardwood trees and abundant ground cover in the
form of logs, rocks, and leaf litter. Habitats include pine-oak forests; mixed
deciduous hardwoods such as oaks, yellow poplar, and American beech; cove
forests; and bottomland forests. The species has also been found in longleaf
pine habitat on rocky hillsides in Louisiana. These salamanders can occupy
relatively drier soils than other woodland salamanders but usually remain
in moist leaf litter and under objects that retain moisture.

BEHAVIOR AND ACTIVITY Southern red-backed salamanders move only a
few linear yards during their entire lives and stay in the same small patch
of forest. Movement is mostly from the surface to underground retreats
during dry periods. They are seldom found on the surface from May to

Southern red-backed salamanders are found in a wide variety of habitat types, most with one or more species of hardwood trees and abundant ground cover.

September but are active during all other months of the year. Juveniles appear in October and adults are first active in November. Like their woodland salamander relatives, these salamanders are territorial.

FOOD AND FEEDING Southern red-backed salamanders eat any animal they can subdue and swallow, including snails, worms, mites, spiders, millipedes, centipedes, pillbugs, thrips, flies, moths, crickets, beetles, ants, and the larvae of many insects. Smaller salamanders eat smaller prey; adults eat both small and large prey.

REPRODUCTION Courtship and mating are terrestrial and take place in winter. Females lay an average of five eggs deep in the ground in June and July and probably remain with them until they hatch in August and September; hatchlings can be found on the surface as early as October. Too few observations have been published to provide a complete picture of the reproductive cycle.

PREDATORS AND DEFENSE No predators have been documented, although undoubtedly small snakes, ground-foraging birds, shrews, and some predatory insects eat these salamanders. Adults and juveniles secrete noxious fluids from skin glands that deter some predators. Individuals become immobile when first exposed, decreasing the likelihood that a visual predator will notice them, and may autotomize the tail when attacked.

CONSERVATION Southern red-backed salamander populations decline when hardwood forests are harvested for timber or cleared for roads, fields, and urban development. This species is not listed as Endangered or Threatened by any state in which it occurs.

TAXONOMY The southern red-backed salamander was first described in 1944 as a subspecies of the eastern red-backed salamander and was elevated to full species status in 1976 on the basis of molecular evidence. No subspecies are recognized.

Wehrle's salamanders are usually found at higher elevations in moist mixed deciduous and coniferous forests, although they sometimes occur on drier, rocky forested hillsides.

How do you identify a Wehrle's salamander?

Wehrle's Salamander

Plethodon wehrlei

DESCRIPTION Wehrle's salamanders are middle-sized members of the woodland salamander group. The body is dark brown or dark gray with small, scattered white spots or brassy flecks on the head, back, and tail. Some individuals have small reddish spots on the shoulders. The sides of the body have abundant bluish white to yellowish spots that often fuse to form blotches or dashes. The belly is uniformly gray, but the throat is white or has white blotches. The webbing on the expanded rear feet reaches almost to the tips of the first two toes.

LARVAE AND JUVENILES Hatchlings and juveniles in southern populations often have pairs of red spots on the shoulders, and those in some areas have paired yellow spots. The yellow spots are retained by adults in isolated populations in Kentucky, Tennessee, and West Virginia.

SIMILAR SPECIES Northern slimy salamanders closely resemble Wehrle's salamanders but have larger and more numerous spots on the back, a gray throat, and 16 costal grooves. Northern gray-cheeked salamanders are uniformly dark gray above and lighter gray below, and have gray patches on the sides of the neck behind the head.

DISTRIBUTION Wehrle's salamanders occur from western New York south through Pennsylvania and West Virginia into the mountains of southwest-

BODY SHAPE
Slender body; large, webbed rear feet

NUMBER OF COSTAL GROOVES
17

BODY COLOR
Dark brown to gray-black

PATTERN
Small white spots concentrated on sides; red or yellow spots on shoulders

BODY SIZE
Adults to 6.5 inches in total length

ern Virginia and North Carolina. Isolated populations occur in eastern Kentucky and northeastern Tennessee.

HABITAT These salamanders are found at elevations above 600 feet, and primarily on high mountains. They prefer mixed deciduous and coniferous forests such as spruce–yellow birch but also occupy dry, forested hillsides where there are rock ledges and flat rocks. They hide under surface objects such as logs, rocks, and leaves but go underground when the surface is too dry. Individuals have also been found in caves.

BEHAVIOR AND ACTIVITY Wehrle's salamanders are active on the surface from March to October at lower elevations, and from May to early October at high elevations. Activity periods are controlled by the timing of rain; salamanders remain underground when temperatures are high and the ground and leaf litter are dry. Like other woodland salamanders, Wehrle's salamanders remain in a small area their entire lives. Adults defend territories—usually including a cover object—on the forest floor from other adults and sometimes other species. They do not migrate but move vertically in response to surface moisture.

FOOD AND FEEDING Adults and juveniles forage for prey on the forest floor and on low vegetation and tree trunks during warm and wet periods. Known prey items include ants, crickets, bugs, beetles, flies, weevils, daddy longlegs, a variety of insect larvae, mites, spiders, springtails, worms, snails, isopods, millipedes, and centipedes.

Did you know?

Some female salamanders are able to store sperm after breeding with a male and produce viable eggs more than a year later.

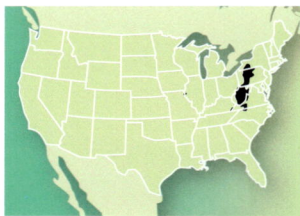

● Wehrle's salamander
Plethodon wehrlei

REPRODUCTION Mating is terrestrial, with males courting females to entice them to pick up sperm caps from spermatophores deposited on the ground. Mating occurs in fall and spring. Females lay 7–24 eggs in underground cavities in hardwood and coniferous forests in March to April. One clutch was found in a cave. Females likely remain with their eggs until they hatch in August to September, as other *Plethodon* females do.

PREDATORS AND DEFENSE The only known predator of Wehrle's salamander is the ringneck snake. Other snakes such as garter snakes, ground-foraging birds, and shrews probably eat them as well.

CONSERVATION Wehrle's salamanders are listed as Threatened in North Carolina. Clearing of hardwood and spruce forests for timber sales, roads, and houses constitutes the most important threat to populations. Changing temperature and rainfall patterns due to climate change may cause contraction of the geographic range.

TAXONOMY Wehrle's salamander was first described in 1917. Some of the geographic variants, including the Dixie Caverns variant in Dixie Caverns, Virginia, and the red-spotted form in southwestern Virginia, were described in the twentieth century as full species but are now recognized as geographic variants.

Adult Wehrle's salamanders in some Kentucky, Tennessee, and West Virginia populations are nearly black with distinct yellow to white spots.

Brassy blotches on the body and tail coupled with a small body size are characteristic of Weller's salamanders, which are found on mountaintops, sometimes at elevations above 5,000 feet.

How do you identify a Weller's salamander?

BODY SHAPE
Small with a thick tail

NUMBER OF COSTAL GROOVES
15–17

BODY COLOR
Black

PATTERN
Abundant brassy markings on the body and tail

BODY SIZE
Adults to 3 inches in total length

Weller's Salamander   *Plethodon welleri*

DESCRIPTION Weller's salamanders are small and dark gray to black with numerous irregular, gold to brassy blotches on the head, back, and tail that often run together to form large patches. The belly varies from black with numerous fine white spots in most of the range to uniformly black in the vicinity of Grandfather Mountain in western North Carolina.

LARVAE AND JUVENILES Hatchlings have several pairs of brassy spots along the back near the head; the body is otherwise black.

SIMILAR SPECIES The lead-backed phase of the eastern red-backed salamander is similar in color but lacks the heavy brassy pattern. Pygmy salamanders have a light stripe down the back and a light eye-to-jaw stripe. Juvenile Yonahlossee salamanders have four to six pairs of reddish spots on the back.

DISTRIBUTION This small salamander occurs on several isolated mountaintops from Mount Rogers and Whitetop Mountain in southwestern Virginia through the Unaka Mountains in the Blue Ridge in northeastern Tennessee and in northwestern North Carolina south as far as Grandfather Mountain.

HABITAT This is a mountaintop species found primarily in spruce-fir forests at elevations above 5,000 feet, and in lower numbers down to about 2,500 feet in moist cove hardwoods. Individuals use logs, rocks, and rock flakes for cover on the forest floor, but spend much of their time under mats of moss and below the surface in leaf and pine needle litter.

BEHAVIOR AND ACTIVITY Weller's salamanders are active from about April or May to October, and are probably inactive near the surface during the cold winter months. These small salamanders defend small territories against intruders, and males attack one another on sight; however, they also aggregate without fighting for long periods during winter.

The loss of the spruce and fir forests in the high elevations of the Blue Ridge to logging and acid rain pollution is a major threat to the long-term persistence of Weller's salamanders.

FOOD AND FEEDING Weller's salamanders forage for insects and other invertebrates beneath the forest floor surface and on the forest floor at night during wet periods. Known prey items of juveniles and adults include beetles, moth larvae, bugs, flies, springtails, mites, spiders, and pseudoscorpions. Individuals eat their own skin when they molt.

REPRODUCTION Mating takes place in spring and fall on the forest floor when the male entices a female—through tail waving and touching and rubbing his chin on her back—to pick up the sperm cap sitting atop a spermatophore. Fertilization is internal.

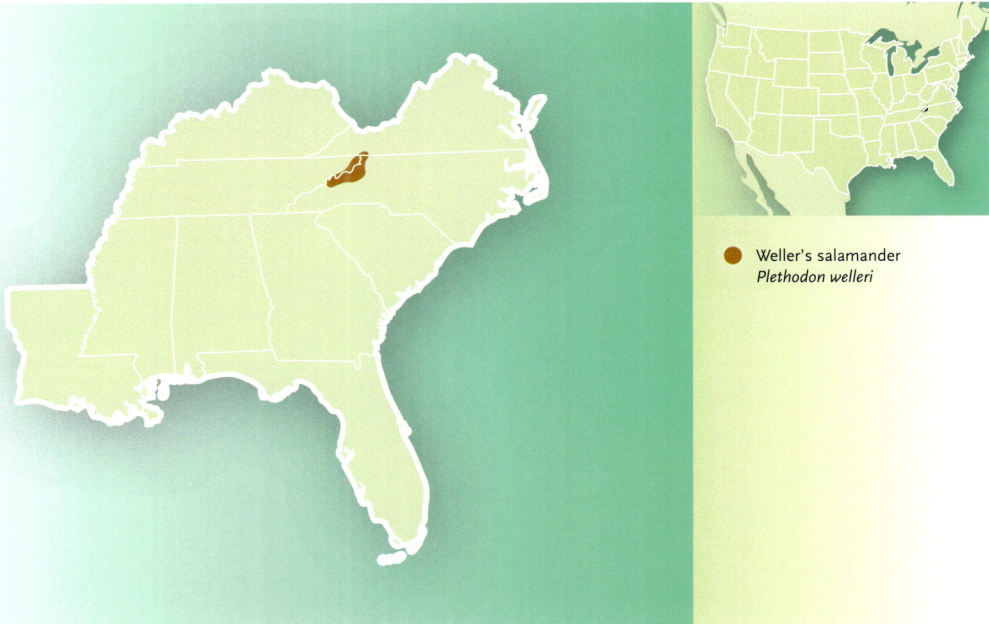

● Weller's salamander
Plethodon welleri

Females lay 4–11 eggs in summer in decomposing spruce or fir logs and under moss mats. The mother stays with the eggs until they hatch in early fall.

PREDATORS AND DEFENSE Predators have not been reported but undoubtedly include birds that forage on the forest floor and possibly small snakes and mammals found at high elevations. Noxious secretions produced by skin glands in the tail are thought to deter some predators. Individuals also become immobile upon discovery, a behavior likely to make visual predators overlook them.

Weller's salamanders forage for insects and for spiders and mites beneath the forest floor surface and on the forest floor at night during wet periods.

CONSERVATION The loss of the spruce and fir forests in the high elevations of the Blue Ridge to logging and acid rain pollution is a major threat to the long-term persistence of this salamander because the forest floor environment is deteriorating. The introduced balsam woolly adelgid has caused serious damage to Fraser fir forests. Increasing temperatures predicted by global climate models may further restrict the range of this species to high mountaintops and could eliminate some lower-elevation populations altogether. Weller's salamander is listed as a Species of Special Concern by North Carolina and is protected by Tennessee as Wildlife in Need of Management.

TAXONOMY This salamander was described in 1931 and named for Worth Hamilton Weller of Cincinnati, Ohio, who died collecting it on Grandfather Mountain in 1931 on a field trip at the age of 18.

Bay Springs Salamander   *Plethodon ainsworthi*

This species is believed to be extinct. No information is available on behavior and activity, reproduction, food and feeding, or predators and defense.

DESCRIPTION Based only on preserved specimens, the Bay Springs salamander is slender, has reduced limbs, and is dark brown or black with no visible pattern.

LARVAE AND JUVENILES Unknown.

SIMILAR SPECIES The Bay Springs salamander resembles other *Plethodon* species within the range but is more slender.

DISTRIBUTION Known only from a location 2 miles south of Bay Springs in Jasper County, Mississippi.

HABITAT The two known specimens were collected in a mixed hardwood habitat with small springs and seepage areas.

CONSERVATION No specific conservation plans are likely to be forthcoming because the species is presumed extinct.

TAXONOMY The Bay Springs salamander was described in 1998 from two specimens collected and preserved in 1964.

How do you identify a Bay Springs salamander?

BODY SHAPE
Slender

NUMBER OF COSTAL GROOVES
16

BODY COLOR
Dark brown or black

PATTERN
None

BODY SIZE
Adults to 4 inches in total length

● Bay Springs salamander
Plethodon ainsworthi

Brass-colored markings on the back and light-colored spots on the sides are characteristic of most Tellico salamanders, which are smaller than adults of most other plethodontid species found within their geographic range.

How do you identify a Tellico salamander?

BODY SHAPE
Slender

NUMBER OF COSTAL GROOVES
16

BODY COLOR
Black

PATTERN
Brass-colored flecks on back, and light spotting on sides

BODY SIZE
Adults to 5.8 inches in total length

Tellico Salamander   *Plethodon aureolus*

DESCRIPTION This small (usually between 4 and 6 inches) black salamander has brass-colored markings on the back and white or yellow spots on the sides. The chin is lighter, and a mental gland is present. The belly is gray and unmarked. The light markings on the body are reduced in some individuals found at higher elevations.

LARVAE AND JUVENILES Juveniles resemble adults.

SIMILAR SPECIES Because of its limited geographic range and small body size, the Tellico salamander is likely to be confused only with the few *Plethodon* species with which it overlaps, most of which are larger. The brass-colored markings are the key character for identification.

DISTRIBUTION The Tellico salamander is known from two counties (Graham and Cherokee) in North Carolina and two (Polk and Monroe) in Tennessee.

HABITAT The species is associated mostly with hardwood forests in lowland areas but is also found at higher elevations above 5,000 feet.

BEHAVIOR AND ACTIVITY Little is known of the general behavior and activity of Tellico salamanders, but they presumably spend most of their time beneath logs, dead leaves, and other ground litter or deeper underground.

Tellico salamander
Plethodon aureolus

The chin of the Tellico salamander is lighter than the dark back and tail.

FOOD AND FEEDING They presumably eat small terrestrial invertebrates.

REPRODUCTION Tellico salamanders have an elaborate terrestrial courtship dance. Nest sites and clutch sizes are unknown, but presumably the eggs are laid underground. Embryos develop in the egg and hatch as miniatures of the adult form, and juveniles are seen above ground in August.

PREDATORS AND DEFENSE Probable predators are small snakes, large spiders, shrews, and birds.

CONSERVATION No serious conservation threats have been identified for this species.

TAXONOMY The Tellico salamander was described in 1984 and has no subspecies. Tellico salamanders hybridize with Jordan's red-cheeked salamanders.

The Pigeon Mountain salamander is found only in Walker and Chattooga counties in northwestern Georgia and can be distinguished from other salamanders in the region by its large size and brown back.

How do you identify a Pigeon Mountain salamander?

BODY SHAPE
Flattened and slender; tail more than half the total length

NUMBER OF COSTAL GROOVES
15 or 16

BODY COLOR
Brown and black

PATTERN
Brown irregular band down back; black tail and sides with white or brassy spots

BODY SIZE
Adults to 7.2 inches in total length

Pigeon Mountain Salamander

Plethodon petraeus

DESCRIPTION These are large salamanders, and females are slightly larger than males. The basic body color of the back and belly is black, but a broad, irregular brown (reddish to olive) band or patch extends the length of the back and onto the sides. The back, sides, limbs, and tail have white or brass-colored spots. Expanded toe tips and an elongated toe on each foot help these salamanders climb rock faces. Males have a noticeable round mental gland.

LARVAE AND JUVENILES Young juveniles have brownish or brass-colored spots on the back that eventually connect to form the pattern characteristic of adults.

SIMILAR SPECIES The brown coloration on the back and expanded toe pads readily distinguish the Pigeon Mountain salamander from other species in the region.

DISTRIBUTION The total range is limited to the eastern side of Pigeon Mountain in Walker and Chattooga counties in northwestern Georgia.

HABITAT The Pigeon Mountain salamander is strongly associated with limestone elements of rocky cliffs, crevices, and cave openings. The surrounding

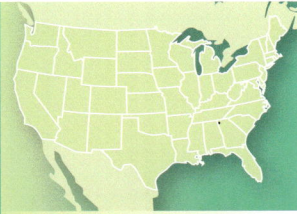

Pigeon Mountain salamander
Plethodon petraeus

woodland habitat is primarily oak-hickory forest.

BEHAVIOR AND ACTIVITY Pigeon Mountain salamanders have physical adaptations that help them climb the sides of rock cliffs and crevices. They are active throughout the year except during the coldest winter temperatures. They can be found under rocks or in caves during most of the year, and in early spring can also be found beneath logs in wooded areas. They are most active above ground on rock surfaces and the forest floor, especially during rains. When conditions are dry, these salamanders will often keep the body concealed in a rock crevice or beneath a rock slab with only the head visible. During dry periods in summer they move into caves or deep crevices.

During dry weather, Pigeon Mountain salamanders will often keep their body concealed in a rock crevice or beneath a rock slab with only the head visible.

FOOD AND FEEDING Primary food items are small invertebrates including terrestrial insects, spiders, and worms.

REPRODUCTION Complete details of the reproductive cycle of the Pigeon Mountain salamander are unknown other than that the breeding season is presumed to be in spring and summer, and eggs are laid later in the sum-

*Species in the
Plethodontidae,
the largest family
of salamanders in
the world, have no
lungs. All terrestrial
salamanders in
this family respire
through their
moist skin.*

The Pigeon Mountain salamander is associated
with limestone topography of rocky cliffs and
crevices that are surrounded by woodland habitat.

mer or in the fall. Clutch sizes of around 17 have been reported. Nests in
nature are unknown, but presumably the eggs are laid underground.

PREDATORS AND DEFENSE Predation has not been observed, but small
snakes, birds, and shrews probably prey on these salamanders.

CONSERVATION The species has a very restricted geographic range and
habitat requirements and is protected by the state of Georgia as a Rare
species.

TAXONOMY The Pigeon Mountain salamander was discovered and recognized as a new species in 1972 and was described in the scientific literature
in 1988. No subspecies are recognized.

This southern Appalachian salamander can be distinguished from most other terrestrial salamanders in its range by its larger size and the presence of white spotting on the body and red pigment on the legs.

How do you identify a southern Appalachian salamander?

BODY SHAPE
Large and stout

NUMBER OF COSTAL GROOVES
16

BODY COLOR
Black

PATTERN
Small white spots on back, larger ones on sides

BODY SIZE
Adults to 8.2 inches in total length

Southern Appalachian Salamander

Plethodon teyahalee

DESCRIPTION These relatively large black salamanders have tiny white spots on the back and tail, and larger white spots on the sides. Some individuals have red spots on their legs. The underside is grayish, with the chin being a lighter gray. The round mental gland is apparent in breeding males.

LARVAE AND JUVENILES Juvenile southern Appalachian salamanders have the same color and markings as adults.

SIMILAR SPECIES Other woodland salamanders found within this species' geographic range and habitats do not have white spotting on the body and red spots on the legs. However, the southern Appalachian salamander hybridizes with Jordan's red-cheeked salamander and possibly other species of woodland salamanders in some areas, and individuals with a mix of color traits are difficult to identify.

DISTRIBUTION The southern Appalachian salamander is restricted to sites in the Blue Ridge physiographic province from southwestern North Carolina to northwestern South Carolina, southeastern Tennessee, and Rabun County in northeastern Georgia. Individuals have been found at elevations approaching 5,000 feet.

HABITAT Adults and juveniles are found in deciduous hardwood forests.

BEHAVIOR AND ACTIVITY Like most other woodland salamanders, this species leads a secretive existence by remaining underground or beneath rocks, logs, dead leaves, and other ground litter most of the time. They are most prevalent above the ground at night during rains.

FOOD AND FEEDING These salamanders eat small terrestrial invertebrates including snails, millipedes, sowbugs, spiders, springtails, and a variety of insects including beetles and ants.

REPRODUCTION Reproduction is terrestrial and presumably similar to *P. jordani*.

Southern Appalachian salamanders are most active above the ground during rains at night.

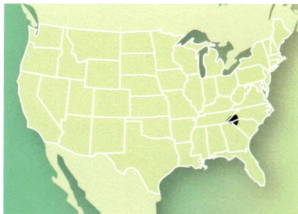

PREDATORS AND DEFENSE No predators have been verified, but small snakes, birds, and shrews probably prey on southern Appalachian salamanders. Aside from leading a clandestine existence and avoiding some predators by coming out only at night, these salamanders produce skin secretions that are mildly toxic or offensive to some predators.

CONSERVATION Concerns for southern Appalachian salamanders apply primarily to local populations that might suffer from degradation of their terrestrial habitat, although the species has been reported to persist in second-growth forests after timbering.

TAXONOMY The southern Appalachian salamander was described in 1950 as *P. jordani teyahalee*, a subspecies of Jordan's red-cheeked salamander, and in 1983 was recognized as a distinct species named *P. oconaluftee*. The latter name did not follow the rules of scientific nomenclature, and the species name was later changed to *P. teyahalee*. No subspecies are recognized.

● southern Appalachian
salamander
Plethodon teyahalee

Southern ravine salamanders inhabit hardwood forests in rocky areas and are more likely to be found beneath rocks or in rock crevices than under logs or leaves. They are found near the surface most often in spring and fall.

How do you identify a ravine salamander?

BODY SHAPE
Slender; short legs

NUMBER OF COSTAL GROOVES
19–23

BODY COLOR
Dark brown to black

PATTERN
Body and tail speckled with silvery or brassy flecks

BODY SIZE
Adults to 5.6 inches in total length

Ravine Salamanders

Northern Ravine Salamander   *Plethodon electromorphus*

Southern Ravine Salamander   *Plethodon richmondi*

DESCRIPTION The body is dark brown or black, and the head, body, and tail are speckled with silver or brass-colored flecks. The underside is similarly dark and has flecks on the chin but not on the belly. The tail constitutes about half the total length, which ranges from about 3 to 5.6 inches. Northern ravine salamanders and southern ravine salamanders are identical externally. Where their ranges overlap in northern Kentucky they can be distinguished only by genetic analysis.

LARVAE AND JUVENILES Recently hatched juveniles are light gray and about 1 inch in total length. Older juveniles develop adult coloration.

SIMILAR SPECIES Several species whose ranges partially overlap that of the southern ravine salamander in the Southeast are superficially similar but can usually be differentiated by body coloration, body shape, or number of costal grooves. The northern zigzag salamander usually has reddish coloration on the belly; Weller's and Wehrle's salamanders have fewer costal grooves; slimy salamanders have a more robust body; and northern gray-cheeked salamanders have gray cheeks.

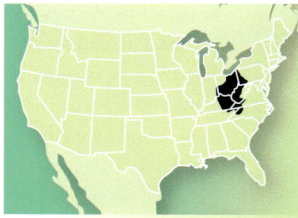

- northern ravine salamander
  Plethodon electromorphus
- southern ravine salamander
  Plethodon richmondi

The belly of ravine salamanders is relatively dark, and the tail is as long as the rest of the body.

DISTRIBUTION The northern ravine salamander ranges from southwestern Pennsylvania through Ohio to southeastern Indiana, and in the Southeast is found only on the northern edge of Kentucky. The southern ravine salamander occurs in southern West Virginia and in parts of four southeastern states: eastern Kentucky, western Virginia, northeastern Tennessee, and northwestern North Carolina.

HABITAT Both species characteristically occur on slopes of hardwood forests with rock outcrops.

BEHAVIOR AND ACTIVITY Ravine salamanders are secretive and often remain below the surface in rock crevices or underground tunnels. They are more likely to be found under rocks than logs and are most likely to be near the surface during fall, winter, and spring.

FOOD AND FEEDING Ravine salamanders feed on a wide variety of small terrestrial invertebrates in all months of the year. Among the major groups eaten are ants, sowbugs, worms, beetles, snails, centipedes, roaches, spiders, and mites.

When threatened by the presence of a predator, ravine salamanders often remain motionless in an attempt to go unobserved.

In the Southeast, the northern ravine salamander is found only in northern Kentucky. It is indistinguishable in appearance from the southern ravine salamander and must be identified on the basis of genetic analysis and geographic location.

REPRODUCTION Mating takes place as early as November and as late as May; the timing depends on local environmental conditions and other unknown factors. Females lay their eggs underground or beneath rocks from April through June, with local variation among populations. Each female lays 5–15 eggs that hatch by early fall, but hatchlings usually remain underground until spring.

PREDATORS AND DEFENSE Ringneck snakes commonly eat ravine salamanders, and other small snakes, large spiders, birds, and small mammals such as shrews probably prey on them as well. One form of defense is to remain motionless in an attempt to go unobserved by a predator.

CONSERVATION The primary threat to these species is loss or degradation of habitat resulting from urban development and forestry activities.

TAXONOMY The southern ravine salamander was described as a species in 1938. The northern ravine salamander, which was originally considered to be the same species as the southern, was described as a separate species in 1999. Neither is partitioned into subspecies.

Webster's salamanders inhabit rocky areas in hardwood forests with heavy ground litter. The species is most likely to be seen on the surface from late spring to early summer.

How do you identify a Webster's salamander?

BODY SHAPE
Small and slender

NUMBER OF COSTAL GROOVES
17–19

BODY COLOR
Brown or reddish brown

PATTERN
Usually a broad stripe down the back and onto the tail

BODY SIZE
Adults to 3.3 inches in total length

Webster's Salamander   *Plethodon websteri*

DESCRIPTION Webster's salamanders are tiny brown salamanders that range in color from reddish to yellowish brown. Some individuals have a stripe down the back to the end of the tail, and a small proportion of the population has no stripe at all. The sides of the stripe are wavy and are usually brighter on the front part of the tail than on the body. The sides have lighter-colored flecks, and the belly is mottled with orange, red, black, and white.

LARVAE AND JUVENILES Recent hatchlings are about 0.5 inch long and resemble adults in coloration.

SIMILAR SPECIES Webster's salamanders can be confused with zigzag salamanders, but the two species' ranges overlap in only a few areas. Because both species are variable in color pattern, the two must be distinguished by genetic analysis.

DISTRIBUTION Webster's salamanders are found from South Carolina to Louisiana, but except for a relatively broad area from western Georgia through east-central Alabama they occur as isolated populations.

HABITAT Hardwood forests in rocky terrain with heavy ground litter of leaves, logs, and rocks are the preferred habitat.

BEHAVIOR AND ACTIVITY These salamanders are locally abundant on the surface beneath ground litter from October to May but virtually disappear deep underground in summer through early fall.

FOOD AND FEEDING Webster's salamanders eat small terrestrial invertebrates such as ants, mites, termites, springtails, spiders, snails, centipedes, flies, and worms.

REPRODUCTION Terrestrial courtship and mating occur from January to March. Females begin laying their clutch of 3–8 eggs underground in June and July. Hatchlings are found above ground in October.

PREDATORS AND DEFENSE Specific predators have not been documented but presumably include ringneck snakes and garter snakes, large spiders, and shrews that occupy similar habitat.

CONSERVATION No generalized conservation threats have been identified for Webster's salamander other than loss of forested habitat. South Carolina lists the species as Endangered, and Louisiana considers it a Species of Special Concern.

TAXONOMY Populations of Webster's salamanders were reported in the scientific literature as early as the 1950s, but as zigzag salamanders. The species was described in 1979, and no subspecies have been recognized.

Did you know?

New species of salamanders continue to be described in the Southeast as DNA analyses differentiate species that look identical. Several new species are expected to be described within the next several years, even as the actual number of salamander populations continues to decline.

Webster's salamander
Plethodon websteri

Most Cheoah Bald salamanders have red pigment on part of their legs, but only about half of the individuals have red pigment on their cheeks.

How do you identify a Cheoah Bald salamander?

BODY SHAPE
Slender

NUMBER OF COSTAL GROOVES
16

BODY COLOR
Uniformly blue-black; usually some red on legs

PATTERN
White to yellow spots, primarily on the sides

BODY SIZE
Adults to 5 inches in total length

Cheoah Bald Salamander   *Plethodon cheoah*

DESCRIPTION The body of these small salamanders is uniformly blue-black except for white to yellow spots on the sides. Usually less than half of the upper surface of the legs is covered by red pigment. The belly is light gray with or without light yellow spots. About half the individuals in any population have a small amount of red pigment on their cheeks.

LARVAE AND JUVENILES Young ones less than 1.25 inches in total length have red spots and brassy flecking on the back, and may have a red V atop the head with the apex pointing toward the snout.

SIMILAR SPECIES Jordan's red-cheeked salamanders do not have red pigment on the upper surface of the legs but are otherwise difficult to distinguish from Cheoah Bald salamanders. Confirmation of identification in the zone where the two species overlap requires molecular analysis.

DISTRIBUTION The species is restricted to Cheoah Bald in Graham and Swain counties in North Carolina and ranges to elevations above 5,000 feet.

HABITAT Higher-elevation forests in rough, rocky terrain are the preferred habitat.

BEHAVIOR AND ACTIVITY Little is known of the year-round activity patterns of the Cheoah Bald salamander other than that individuals move underground during the coldest and driest periods. When on the surface, they stay under rocks and logs.

FOOD AND FEEDING Cheoah Bald salamanders are presumed to eat an array of terrestrial invertebrates.

REPRODUCTION Following courtship and mating on the ground surface, females are presumed to retreat underground to lay their eggs, and to stay with the eggs until they hatch. Clutch size and incubation period are unknown.

PREDATORS AND DEFENSE Probable predators are small snakes, large spiders, and shrews.

CONSERVATION No general conservation threats are imminent for this species, which is not protected in its limited geographic range in North Carolina.

TAXONOMY Originally presumed to be a local population of Jordan's red-cheeked salamander, the Cheoah Bald salamander was described as a new species in 2000 and has no recognized subspecies. Cheoah Bald salamanders hybridize with southern Appalachian salamanders.

Cheoah Bald salamanders have a limited geographic range, being restricted to two counties in North Carolina. The species can be found at elevations above 5,000 feet.

Northern zigzag salamanders look exactly like southern zigzag salamanders, and some salamander experts consider them to be the same species. In the Southeast, northern zigzags are found only in Kentucky.

How do you identify a zigzag salamander?

BODY SHAPE
Small and stout

NUMBER OF COSTAL GROOVES
17–19, usually 18

BODY COLOR
Brown or dark gray

PATTERN
Reddish to yellowish stripe on back and tail, or no stripe

BODY SIZE
3.5 inches in total length

Zigzag Salamanders

Northern Zigzag Salamander *Plethodon dorsalis*

Southern Zigzag Salamander *Plethodon ventralis*

DESCRIPTION Zigzag salamanders are small, brownish or dark gray salamanders that vary considerably in color and pattern. The color ranges from reddish to yellowish brown, and individuals either have a wavy central stripe down the back and onto the tail or no stripe at all. The sides of the stripe are scalloped toward the front of the body. The dark sides have light-colored flecks. The belly is a mix of orange, red, black, and white. A pale red shoulder patch is present on most individuals in middle Tennessee.

LARVAE AND JUVENILES Young juveniles about 2 inches long appear on the surface in September. Although this terrestrial salamander bypasses the aquatic larval stage and develops completely within the egg, remnant gill stubs are occasionally present in hatchlings. Juveniles resemble adults. The two zigzag species are indistinguishable in appearance but can usually be identified by geographic location.

SIMILAR SPECIES The southern zigzag salamander's geographic range overlaps that of Webster's salamander only in Alabama (Jefferson County). The southern red-backed salamander looks similar but has more costal grooves (18–22).

Individual zigzag salamanders can have a central stripe (see previous page) or have no stripe like this one.

Zigzag salamanders have bellies that are a mix of orange, red, black, and white.

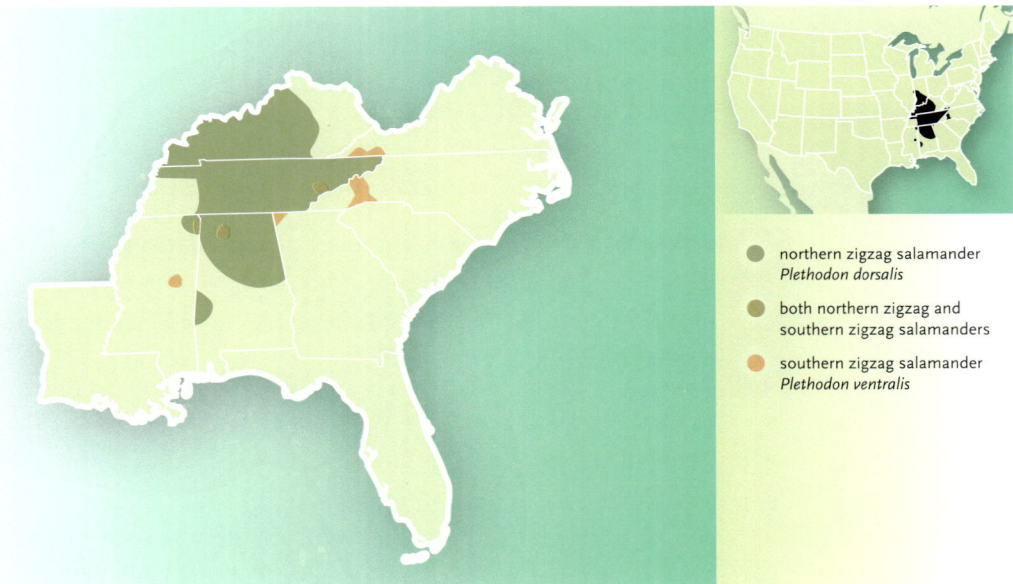

● northern zigzag salamander
Plethodon dorsalis

● both northern zigzag and southern zigzag salamanders

● southern zigzag salamander
Plethodon ventralis

DISTRIBUTION The inclusive geographic range of the northern and southern zigzag salamanders is from southern Illinois and Indiana south through Kentucky to Mississippi; the two species overlap in Kentucky. The southern zigzag salamander is restricted to the Southeast, occurring in locations in western Virginia and North Carolina, eastern Kentucky and Tennessee, northwestern Georgia, northern Alabama, and central Mississippi.

HABITAT The primary habitat of zigzag salamanders is hardwood forests, often in rocky areas. During the warm months they are common inhabitants of subterranean passages and can often be found in crevices in cave floors and walls.

BEHAVIOR AND ACTIVITY During fall and winter, zigzag salamanders commonly stay beneath the ground litter of dead leaves, logs, and rocks during the day and walk around on the ground, roots, bases of shrubs, and rocks while foraging at night during wet weather. With the onset of warm, dry weather they retreat underground or into caves, where they remain until the cooler and damper conditions of fall return.

FOOD AND FEEDING Zigzag salamanders eat spiders and beetles, and probably any other small terrestrial invertebrates they can catch and swallow.

REPRODUCTION Courtship and mating take place from fall to spring, but the specific timing of reproduction varies among populations on the basis of winter temperature and moisture conditions. Each female lays 1–10 eggs and stays with them during development, which takes about 3 months. Very few nests have been found, and all were in caves.

PREDATORS AND DEFENSE Screech owls and ringneck snakes are known predators, and garter snakes, large spiders, and shrews and other small mammals probably eat zigzag salamanders as well.

CONSERVATION Local populations of zigzag salamanders can be reduced in size or eliminated by improper timbering or other land-clearing activities. North Carolina lists the southern zigzag salamander as a Species of Special Concern.

TAXONOMY The original description of *Plethodon dorsalis* in 1889 included the southern and northern forms as a single species. The southern zigzag salamander was described in 1997 as a distinct species on the basis of genetic evidence. Northern and southern zigzag salamanders interbreed naturally in parts of Kentucky, and some salamander authorities do not consider the two to be separate species. The Ozark zigzag salamander (*Plethodon angusticlavius*) of the Ozark Plateau was once considered a subspecies of *Plethodon dorsalis*, and Webster's salamander was included in the zigzag salamander species until 1979.

A female zigzag salamander lays up to 10 eggs and stays with them during development, which may take 3 months.

Zigzag salamanders bypass the aquatic larval stage and develop completely within the egg in the terrestrial habitat.

Red-legged salamanders have a dark blue or black back, typically with bright red markings on the legs.

Red-legged Salamander   *Plethodon shermani*

DESCRIPTION These large salamanders have a gunmetal blue–black back and tail. The belly is bluish gray. The bright red markings on the legs are a signature trait of the species, but red coloration is reduced or absent on some individuals. Females get larger than males.

LARVAE AND JUVENILES By the time they appear on the surface, juveniles resemble adults.

SIMILAR SPECIES Red-legged salamanders hybridize with southern Appalachian salamanders at several locations, and genetic analysis may be required to determine the two species.

DISTRIBUTION The species is found in southwestern North Carolina in the Nantahala Mountains and in Towns County in northwestern Georgia, and has been reported in Monroe County in southeastern Tennessee.

HABITAT Red-legged salamanders occupy wet, cool mountain hardwood forests with rhododendron.

BEHAVIOR AND ACTIVITY These woodland salamanders remain beneath logs, fallen bark, and rocks during the day and may emerge onto the surface at night or during rains. During cold winters they retreat deeper underground.

How do you identify a red-legged salamander?

BODY SHAPE
Large and slender; protruding eyes

NUMBER OF COSTAL GROOVES
15 or 16

BODY COLOR
Bluish black

PATTERN
Solid back with red blotches on legs

BODY SIZE
Adults to 6.7 inches in total length

red-legged salamander
Plethodon shermani

FOOD AND FEEDING Red-legged salamanders eat a wide array of terrestrial invertebrates including many insects (ants, beetles, flies, wasps, aphids, and craneflies), spiders, springtails, millipedes, and earthworms.

REPRODUCTION Courtship and mating are terrestrial, and eggs are presumably laid underground. Clutch size and incubation period are unknown.

PREDATORS AND DEFENSE Predators presumably include small terrestrial snakes, large salamanders, shrews, and birds, but nothing has been published on the subject.

Although red legs are characteristic of most red-legged salamanders, some individuals lack the red coloration and have black legs.

CONSERVATION Despite the relatively small geographic range, the habitat of red-legged salamanders is mostly in national forests and does not appear to be threatened by development or timbering activities.

TAXONOMY The red-legged salamander was described as a distinct species in 1906. Some herpetologists later classified it as a subspecies of slimy salamander before it was placed in the *Plethodon jordani* complex as a subspecies. It was reinstated as a full species in 2000.

The blue-black back and tail, bluish gray belly, and distinctive pink or red cheeks are trademark characters of the Jordan's red-cheeked salamander, although the red coloration is absent in some specimens.

How do you identify a Jordan's red-cheeked salamander?

BODY SHAPE
Large and slender; protruding eyes

NUMBER OF COSTAL GROOVES
15 or 16

BODY COLOR
Black

PATTERN
Solid back with reddish cheeks

BODY SIZE
Adults to 6.7 inches in total length

Jordan's Red-cheeked Salamander

Plethodon jordani

DESCRIPTION These large salamanders have a black back and tail, bluish gray belly, and pink or red cheeks. The head is sometimes brownish. The red coloration is sometimes reduced or, rarely, absent entirely.

LARVAE AND JUVENILES Juveniles remain underground for almost a year after hatching and appear on the forest floor in May. They resemble the adults.

SIMILAR SPECIES The red cheeks and black body readily distinguish the Jordan's red-cheeked salamander from most other salamanders within its geographic range. Jordan's red-cheeked salamanders can hybridize with southern Appalachian salamanders, and differentiating between the two species requires genetic analysis.

DISTRIBUTION Jordan's red-cheeked salamanders are found in Tennessee and North Carolina within the boundaries of Great Smoky Mountains National Park.

HABITAT Slopes and ridges with hardwoods and spruce-fir forests with abundant ground cover of mosses, logs, and rocks are preferred habitats.

Jordan's red-cheeked salamanders crawl around on the surface at night and during rains but remain hidden beneath logs, moss, and rocks during the day.

BEHAVIOR AND ACTIVITY These salamanders remain hidden beneath logs, moss, and rocks during the day and crawl around on the surface at night and during rains. They retreat deeper underground when the surface is dry.

FOOD AND FEEDING Numerous invertebrates have been recorded in the diet of Jordan's red-cheeked salamanders, with expected variations from season to season and among geographic locations. Among the major groups the salamanders prey on are millipedes, insect larvae, worms, snails, beetles, spiders, springtails, flies, and ants.

REPRODUCTION All aspects of reproduction are terrestrial. Clutch size is unknown; eggs are presumably laid in underground burrows.

Jordan's red-cheeked salamanders retreat deeper underground during extended dry periods.

PREDATORS AND DEFENSE Black-bellied salamanders, spring salamanders, eastern garter snakes, and short-tailed shrews eat Jordan's red-cheeked salamanders. They are probably also preyed on by ringneck snakes and birds, although their noxious skin secretions may repel some predators.

CONSERVATION Jordan's red-cheeked salamanders occur within the Great Smoky Mountains National Park and are not threatened from a conservation perspective. However, global warming could be a future threat because this species depends on cool, moist conditions.

TAXONOMY The Jordan's red-cheeked salamander was described as a species in 1901. It is sometimes known as Jordan's salamander and may be called the red-necked salamander in older field guides.

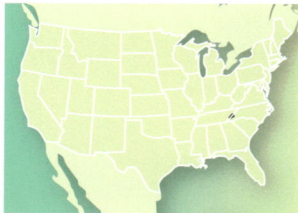

Jordan's red-cheeked salamander
Plethodon jordani

Cow Knob salamanders are found in mixed hardwood forests at high elevations, most commonly on cool, north-facing slopes.

Cow Knob Salamander

Plethodon punctatus

DESCRIPTION Cow Knob salamanders are large and grayish black with a profusion of irregular white to cream-colored spots on the body, legs, and tail; the spots are more abundant on the sides than on the back. The throat is light to pinkish gray, and the belly is dark gray with small cream-colored spots. The Cow Knob salamander superficially resembles a slimy salamander but has thinner, more delicate-looking legs. Its head is flatter as well, and the eyes are more protruding. The top of the head is often slightly purplish relative to the gray-black body.

LARVAE AND JUVENILES Juvenile Cow Knob salamanders appear on the surface in September and are similar to the adults in appearance, although they have fewer light spots.

SIMILAR SPECIES Cow Knob salamanders superficially resemble two species of slimy salamanders that occur in the area. Northern slimy salamanders are patterned with brassy to silvery spots. White-spotted slimy salamanders have large, irregular white spots. Both have 16 or fewer costal grooves. The Cow Knob salamander is closely related to Wehrle's salamander and, although similar in appearance, lacks the latter's red pigment and brassy flecking. These two species apparently do not overlap geographically.

How do you identify a Cow Knob salamander?

BODY SHAPE
Slender

NUMBER OF COSTAL GROOVES
17 or 18

BODY COLOR
Dark gray to black

PATTERN
Body, legs, and tail with white or cream-colored spots

BODY SIZE
6.2 inches in total length

DISTRIBUTION Cow Knob salamanders occur on Shenandoah Mountain and Great North Mountain in the George Washington National Forest in Augusta, Rockingham, and Shenandoah counties in Virginia; and Pendleton and Hardy counties in West Virginia.

HABITAT This species is restricted to mixed hardwood forests in rocky areas at higher elevations (above 2,500 feet), and most populations occupy cooler, north-facing slopes. These salamanders are most common in areas of weathered talus rock where soil and leaf litter are also plentiful. The largest populations have been found in association with old hemlock forest stands.

Cow Knob salamanders are grayish black with a profusion of irregular white to cream-colored spots on the body, legs, and tail.

BEHAVIOR AND ACTIVITY Cow Knob salamanders are generally active above ground in May and June, and again—but not as abundantly—in September and October when environmental conditions are wet and cool. They remain under logs and rock cover during the day but are active on the surface at night as they forage on the ground, rocks, and the base of trees. They retreat underground in July and August when surface conditions are dry.

FOOD AND FEEDING Cow Knob salamanders eat a variety of small invertebrates, including ants, springtails, beetles, flies, millipedes, centipedes, spiders, and mites. Adults consume larger and more varied prey than juveniles.

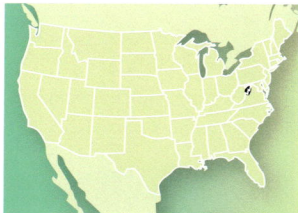

Cow Knob salamander
Plethodon punctatus

REPRODUCTION Cow Knob salamanders are presumed to breed terrestrially and to lay eggs underground in the summer. No nests have been found or described. Each female lays 7–16 eggs and probably remains with them until hatching, as is typical of most members of the genus.

PREDATORS AND DEFENSE Small snakes, birds, shrews, ground-feeding birds, and possibly raccoons and opossums are probably this salamander's primary predators, although no specific observations have been published. Toxic secretions produced by glands in the tail are presumably effective at discouraging some predators.

CONSERVATION The very restricted geographic range of the Cow Knob salamander makes the species potentially vulnerable to timber operations. The regional decline in hemlock trees because of the introduced hemlock woolly adelgid, loss of the canopy because of the gypsy moth, and climate change threaten the habitat. The U.S. Fish and Wildlife Service lists the Cow Know salamander as a Species at Risk. Both Virginia and West Virginia list it as a Species of Special Concern, which means that all management activities in the Shenandoah Mountain Crest Special Biological Area in the George Washington National Forest, where much of its range lies, must be reviewed for their potential impact on the salamander before implementation.

TAXONOMY The Cow Knob salamander was officially described as a species in 1972. No subspecies are recognized. This species has also been called the white-spotted salamander.

The very restricted geographic range of the Cow Knob salamander makes the species potentially vulnerable to timber operations.

A researcher from Virginia Tech releases a protected hellbender into an Appalachian river in southwestern Virginia.

People and Salamanders

WHAT IS A HERPETOLOGIST?

A scientist who studies reptiles and amphibians (including salamanders) is called a herpetologist. Many zookeepers, pet hobbyists, and wildlife managers also study reptiles and amphibians, often with interests in specific groups.

Why Do Herpetologists Study Salamanders?

Salamanders can be studied in classrooms, in laboratories, and in the wild. One of the first animals premedical students have traditionally studied in a comparative anatomy course is a large salamander—the mudpuppy (*Necturus maculosus*). Students dissect preserved mudpuppy specimens to examine the circulatory, nervous, respiratory, muscular, and skeletal systems, which have the basic design of these systems in humans. Biological supply houses have provided hundreds of thousands of mudpuppy specimens for anatomy courses.

Salamander species that are easily maintained in captivity have been used extensively in laboratory experiments to address questions related to how salamanders defend themselves from predators, whether they engage in competitive behavior toward other salamanders, and whether they can

These marbled salamanders were collected overnight in a single bucket at a drift fence by researchers at the Savannah River Ecology Laboratory in Aiken, South Carolina.

Although in different genera, the Jordan's red-cheeked salamander (top) and imitator salamander (bottom) can be difficult to tell apart.

The Shenandoah salamander (left) and red-backed salamander (right), both members of the genus Plethodon, can be difficult to distinguish. Note the width of the dorsal stripe.

recognize relatives as individuals. Larval and adult tiger salamanders have been used for decades to address developmental biology questions related to regeneration of body parts.

Some conservation biologists focus their research on salamanders because many species in the United States and elsewhere are currently at risk. Salamander populations that are sensitive to forestry or mining activities or to urban development are good indicator species; that is, their status indicates the health of the overall ecosystem.

How Do Herpetologists Study Salamanders?

Herpetologists who study salamander ecology and behavior in the field use many techniques; some are standard for all studies, and some are designed for specific situations. The basic biological information gained from such studies is often applicable to issues related to conservation of particular salamander species or habitats. Among the typical measurements herpetologists record in field studies of salamanders are body, or snout–vent, length (the distance from the tip of the snout to the vent—the posterior margin of the cloacal opening); tail length; and body weight. Date, time, and location of capture are routinely recorded too, as is the sex of each individual when it can be determined. If genetic studies are being conducted on a salamander population, tissue samples of blood or muscle (such as a toe or the tip of the tail) may be taken for DNA analyses.

Many herpetological field studies use the mark-release-recapture method to assess population size, determine individual growth rates, and estimate survivorship patterns. Salamanders can be marked for individual identification by clipping toes according to a coding scheme that gives unique combinations. This widespread, although somewhat controversial, technique has proven very effective in long-term studies of salamanders. The elastomer technique is a more modern method that has also been used with success in field studies. An elastomer is a silicone-based, fluorescent material that can be injected beneath the skin of a salamander and seen with an ultraviolet light source. Known as the "visible implant elastomer" (VIE) tagging system, the technique uses different colors implanted at different points on the body, giving hundreds of possible combinations so that individuals can be recognized. Elastomer implants are presumed to last for the life of the marked animal and are therefore useful in long-term mark-release-recapture studies. Digital photography allows individual salamanders to be identified using body color patterns, which are permanent in adult marbled salamanders and newts, and presumably in other species as well. This approach is similar to analyzing a fingerprint record. Another approach for individually marking animals that has had success with larger salamanders is the use of passive integrated transponder (PIT) tags—tiny, glass-encapsulated electronic devices that can be injected into the body cavity of the salamander. A special hand-held reader determines the specific code emitted by the PIT tag.

Several techniques are used to collect salamanders—both aquatic and terrestrial forms—in the wild. If the season is right, the fundamental approach of turning over logs, rocks, and other ground litter is effective for finding the adult stage of most species. Herpetologists capitalize on salamanders' tendency to hide beneath ground cover by placing sheets of plywood known as coverboards in areas where terrestrial salamanders

All salamanders have teeth, and some of the larger species can give a painful bite (amphiuma, top; hellbender, bottom). Such injuries are rare, however, and nearly always occur after someone picks up the animal.

Several techniques are used to collect salamanders in the wild, including small funnel traps commonly used to capture minnows (top right) and active dip netting (bottom right) in the water. Common techniques for capturing salamanders on land capitalize on the tendency of salamanders to hide beneath ground cover by placing sheets of plywood known as coverboards on the ground (top left) with vertical drift fences with pitfall traps alongside them (bottom left).

are known to occur. The coverboard technique is highly effective for collecting some species of ground-dwelling salamanders.

Vertical drift fences (which resemble the silt fences used in construction areas) with pitfall traps alongside them are a highly successful method for capturing seasonal pond–breeding salamanders. A salamander that encounters a drift fence will generally turn left or right and follow the fence, and will eventually fall into a pitfall—a bucket sunk into the ground next to the fence. A common and effective practice is to encircle an aquatic breeding site so that individual salamanders moving from the terrestrial habitat will encounter the fence. Literally thousands of marbled salamanders (*Ambystoma opacum*) and mole salamanders (*A. talpoideum*) have been captured in a single night at the Savannah River Ecology Laboratory in South Carolina with this technique. Such exceptional capture records can occur in the fall when adults come in to breed or in the spring and summer when recently metamorphosed larvae are emigrating from the wetland.

A good way to examine aquatic salamander larvae on a field trip is to place the animals in a small plastic bag filled with water.

Radiotransmitters that can be set at a particular frequency and tracked with an antenna and a receiver have been used in many wildlife studies. The development of transmitter units small enough to be implanted internally has been of great value in determining the distances individual salamanders travel into the surrounding uplands from wetlands. Pond-breeding tiger salamanders, for example, have been followed for weeks.

The study technique selected will depend on several factors, including the goal of the study, the size of the animal, and the impact a particular technique will have on the habitat and the other animals that share it. Invasive techniques are generally used only when noninvasive ones will not suffice.

A herpetologist who wants to catch salamanders for research or for a classroom project must be aware of and must follow all of the state and federal laws that have been passed to protect native wildlife. A researcher in any southeastern state must obtain a scientific collecting permit in order to study salamanders in the wild, regardless of whether the animals will be taken to the laboratory or simply marked and measured in the field. For example, a permit from the U.S. Fish and Wildlife Service is required for catching and measuring either species of flatwoods salamanders, which are protected under the U.S. Endangered Species Act. Collecting protected species for captive research purposes is not permitted, although if the decline continues, captive breeding programs may be necessary in the future.

Each state in the Southeast has wildlife regulations that differ in a variety of ways, including which species are protected, how many a person can possess without a permit, and which species cannot be brought into the state

Salamanders can usually be found under rocks in or along small streams in the Appalachians. They cannot be removed from state or national parks without a permit.

Salamanders make excellent photographic subjects and are popular among both amateur and professional photographers.

or sold there. Many states also have regulations that limit or even prevent the release of salamanders that have been held in captivity. Although obtaining permits may require time-consuming paperwork and may even involve purchasing a scientific collecting permit, most state and federal regulations designed to protect native nongame wildlife such as salamanders are meant to preserve the well-being of all the species in the region.

KEEPING SALAMANDERS AS PETS

Salamanders do not have the widespread popularity as pets that many reptiles or even some frogs do. A few species, such as the fire-bellied newts (genus *Cynops*) and tiger salamanders (*A. tigrinum*), can be found commonly in the commercial pet trade, and other species sometimes become available. In fact, practically any species of salamander can be raised in captivity as long as proper housing, moisture, and food are provided. As is the case with all native wildlife species, state and federal regulations must be considered before taking a salamander from the wild because some species are protected. A person who plans to keep any wild animal in captivity must also consider ethical and conservation-related issues.

Practical and Ethical Considerations

Most people who decide they want a salamander as a pet will probably want to catch their own. Before doing that, they must consider how the removal of individuals from the wild might affect the natural populations and also whether the removed individuals will do well in captivity. Opinions among herpetologists vary considerably about the capture and removal of salamanders from the wild. Some think that removing one or a few salamanders (or other herpetofauna that are not threatened or endangered) as pets will have no long-term effect on a local population. In this view, the negative environmental consequences are minimal or nonexistent, and caring for the animals and learning their habits will enhance the pet owner's respect for both the animals and their environment. Other herpetologists believe that no wild salamander should ever be taken as a pet.

Removing any animal from the wild to keep as a pet carries responsibility along with enjoyment. The pet owner becomes responsible for the animal's well-being for the rest of its life, which in some salamanders may be more than 10 or even 20 years. Someone wanting a salamander for a pet must therefore consider what would happen if it should become impossible to keep the animal. It is illegal in some states to release a pet animal back into the wild after it has been kept in captivity. If such release is legal, other questions arise concerning how local salamanders could be affected by a released pet entering the population and how the individual pet itself will fare in the wild. Will a salamander fed as a pet know how to find its own food? Does a pet salamander have a communicable disease that has not expressed itself in captivity but might be transmitted to salamanders in a wild population? Is the area where the pet is released appropriate habitat, and is the animal being released at the proper season? Is there the chance that the individual could hybridize with local salamanders of a different but closely related species and interfere with the local population dynamics and genetics? All of these issues need to be addressed by anyone who decides to have a salamander as a pet.

Salamander Husbandry

Salamanders are not widely promoted in the commercial pet trade, and the life history and ecology are poorly known for most species. Permanently aquatic species such as sirens, mudpuppies, and waterdogs can usually be kept under aquarium conditions that would be suitable for freshwater fishes. Most terrestrial species will need a moist terrarium with mosses, flat rocks, and bark for hiding. Heating is not the problem with salamanders that it can be with pet reptiles. Cool, humid conditions are generally suitable for most salamanders, and sunlight or ultraviolet lamps are not necessary. Most salamander species have not been bred successfully in captivity, in part because so many of them spend most of their lives in one habitat and breed in another.

All salamanders are carnivorous, both as larvae and as adults. Depending on their life stage, they eat primarily aquatic or terrestrial invertebrates. Most will eat crickets, mealworms, and earthworms that can be found at bait stores and pet shops. Dietary supplements may be required for long-term care, although specific information is lacking.

Did you know?

Some of the salamanders discovered in the mountains of Central America in the late 1900s and early 2000s became extinct because of habitat destruction before the scientific publications describing them even appeared in print.

Salamanders that breed in ponds live most of their lives underground in nearby forests and often move up to a thousand yards to find suitable refugia.

Many salamanders are killed by "urban wildlife" such as raccoons.

A Conservation Vision for Salamanders

Salamanders belong to an ancient lineage more than 200 million years old. The ancestors of modern salamanders survived the dinosaur extinctions. In other words, the salamanders we see today are the descendents of animals that successfully reproduced and left offspring through many millennia, surviving changing climates and escaping predators of all types. Copies of many of the same genes that occurred in those salamanders from eons ago still exist today in the cells of the salamander you find under a log or see swimming in a pool. Unfortunately, scientific studies begun several decades ago are showing that many of the prehistoric lineages now represented by modern species will not survive human impacts to their environments. Several species have already been declared extinct, and many more will follow.

Five massive extinction episodes have occurred in the history of the planet, all as a result of natural changes or disasters. Are we to be the cause of the sixth great extinction? Some scientists think so—and think that it will happen in our lifetimes or those of our children. The process has already begun. What can we do about it? Should we protect these animals on moral or practical grounds? One moral argument is that we do not have the right to cause the extinction of another life form, especially one that does not harm us in any way. Another is that an encounter with a

Children can be taught to appreciate salamanders, especially when other children are the teachers. Tiger salamanders are attractive and harmless.

colorful salamander is a soul-satisfying experience that connects us to the spirit in nature. Practically speaking, chemical derivatives from the secretions of salamander glands might someday create a wonder drug that cures certain forms of cancer or other diseases. Salamanders already perform valuable ecological services that benefit humans, such as consumption of larval mosquitoes.

The southeastern landscape is changing, for all of its inhabitants. The old farms, plantations, and timberlands are gone, and urban sprawl is eating up countless square miles. Human populations are growing in all states in the region; with them come road networks, impoundments for drinking water, sewer systems, stream channelizations, and more extensive recreational use of natural areas. Loss of habitat is the primary and most widespread effect of human activities—and it is directly responsible for the loss of salamander populations. Some 17 million acres have been converted to urban and suburban environments in the United States since the 1950s. The human population of Florida increased from 2.7 million in 1950 to 18 million in 2006, and most of those people live in expanding population centers.

Intensive urbanization and untold miles of roads not only destroy natural habitats, they also fragment them into smaller and smaller patches. Salamanders living in these patches are unable to migrate to other fragments of their original population through the matrix of barriers and inhospitable land. Many of those that try are killed while crossing roads. The resulting smaller and smaller populations are more vulnerable to inbreeding, catastrophic weather events, and disease. Small populations in small patches are killed by feral cats and raccoons that are subsidized by human resources. Thus, urbanization and road construction compound a host of other factors that collectively stack the deck against any salamander population living in

Urban sprawl has destroyed old farms, plantations, wetlands, and timberlands throughout the Southeast.

Salamanders living in isolated patches of habitat cannot migrate to other fragments across the matrix of barriers and inhospitable lands that humans have created.

a habitat fragment. Similar scenarios also occur in agricultural and timber lands. The striped newt, for example, has declined in places where it was once abundant as a result of loss of forest cover, drainage of wetlands, introduction of fish, and fire suppression. Only 15 populations were known to exist in the first decade of the twenty-first century.

Other important causes of salamander population declines include pollution from agriculture, industrial, and private lawns; emergent infectious diseases that were not known a decade ago; and invasive species. Chemicals in agricultural fertilizers, such as atrazine, have been shown to cause mortality and deformities in salamanders and frogs. Chytrid fungus, ranaviruses, and other disease-causing organisms unknown to science until near the end of the twentieth century are now implicated as the cause of amphibian population declines in many areas. Populations of spotted salamanders in western North Carolina have been severely reduced by disease. Several species of earthworms brought to this region have displaced native earthworms and altered the soil humus–leaf litter characteristics of forests, rendering this microhabitat inhospitable for many of the woodland salamanders in the genus *Plethodon*. Invasive plants such as cogongrass also alter soil structure.

Global climate change is predicted to cause warmer temperatures in the Southeast along with longer periods of drought. Warmer temperatures are expected to drive mountain salamanders to higher and higher elevations. If that happens, the highest-elevation species, such as the pygmy salamander, will have no place to go. Longer periods of drought will reduce water levels in ponds and may eliminate the seasonal wetlands and other ephemeral pools mole salamanders require for breeding.

Did you know?

Diseases are emerging in increasing numbers of salamander species and populations in the Southeast and worldwide. The chytrid fungus, for example, has caused declines of salamanders in the American West and extinctions of species in the tropics.

A female spotted salamander killed during the breeding season by a raccoon

We have it within our power to reduce these human-caused threats and find ways to ameliorate their effects on salamanders. Initial efforts to protect amphibians from extinction started with the passage of federal and state regulations, the most effective being the U.S. Endangered Species Act. Such laws safeguard against commercial collection and killing but, except for the federal act, provide only marginal protection against habitat loss. In addition to the U.S. Fish and Wildlife Service, which administers the endangered and threatened species list, the wildlife agencies within each state have their own lists of species that warrant protection. Efforts to protect salamander species on federal and state lists, and adding other species to such lists, will likely be a part of the salamander conservation tool box for some time to come.

We cannot rely on government regulations to protect all southeastern salamanders. Each of us can and should play a role in habitat conservation and protection. The focus should be on habitats where these animals live, because without viable habitat all other protection is futile. Habitat management guidelines exist for all southeastern habitats (see Further Reading) and are readily available to everyone. Landowners can manage the habitats on their property to protect species already there or even attract local species by finding ways to connect isolated fragments of habitat. Protecting natural wetlands of all types from pollution, ditching, and filling benefits both salamanders and frogs; however, it is also essential to protect and manage the surrounding upland—usually forested—habitat. Salamanders may breed in ponds, but many live most of their lives underground in the forest and often move up to a thousand yards away to find suitable refugia. Thus, the combined wetland and upland habitat unit should be the target of conservation efforts. Learn what types of habitats are used by salamanders on your land. Work with knowledgeable persons to find ways to make your property more salamander friendly.

Habitat restoration and creation are increasingly being used to ensure that local populations of species do not go extinct. Effective ways to restore natural wetlands are in place (see Further Reading). The Natural Resources Conservation Service (NRCS) branch in every county in the Southeast is staffed with people well versed in wetland restoration. In some cases, it may be advisable to relocate a number of salamanders from a large population to another area to reestablish a population that has been lost. Opportunities to modify degraded lands in ways that result in establishment and restoration of natural habitats are everywhere. Many of these efforts are still trial and error, but we must start somewhere to restore places that once supported salamander populations. Information may be found on websites or in your local college or NRCS office.

The Southeast is a hotspot of salamander diversity. There are ample opportunities to learn about the 102—well, 101 since the Bay Springs salamander is now extinct—salamanders living in this region. These diverse and beautiful amphibians cannot be saved without your help. Your contributions to efforts to conserve and manage habitats, to support educational programs, and to educate your local land use planning boards—and indeed all of your elected officials—are a good way to be involved as concerned citizens. If you own land, show your neighbors the value of protecting and managing natural habitats. Conservation of these ancient amphibian lineages requires interactions among partners from all walks of life. And that means you.

Habitat restoration and creation are increasingly being used to preserve local populations of salamanders. This created wetland is in eastern Tennessee.

What kinds of salamanders are found in your state?

COMMON NAME	LA	MS	AL	GA	FL	SC	NC	VA	TN	KY
AQUATIC SALAMANDERS										
One-toed Amphiuma		●	●	●	●					
Two-toed Amphiuma	●	●	●	●	●	●	●	●		
Three-toed Amphiuma	●	●	●						●	●
Hellbender		●	●	●		●	●	●	●	●
Shovel-nosed Salamander				●		●	●	●	●	
Many-lined Salamander				●	●		●	●		
Black Warrior River Waterdog			●							
Gulf Coast Waterdog	●	●	●							
Neuse River Waterdog							●			
Dwarf Waterdog				●			●	●		
Mudpuppy	●	●		●			●	●	●	●
Southern Dwarf Siren					●					
Northern Dwarf Siren				●	●	●				
Lesser Siren	●	●	●	●	●	●	●	●	●	●
Greater Siren			●	●	●	●	●	●		
AQUATIC CAVE SALAMANDERS										
Georgia Blind Salamander				●	●					
Tennessee Cave Salamander			●	●					●	
Berry Cave Salamander									●	
STREAM AND SEEP SALAMANDERS										
Streamside Salamander									●	●
Seepage Salamander			●	●		●	●		●	
Southern Dusky Salamander	●	●	●	●	●	●	●	●		
Imitator Salamander							●		●	
Black Mountain Salamander								●	●	●
Black-bellied Salamander				●		●	●	●	●	
Dwarf Black-bellied Salamander				●						
Ocoee Salamander			●	●		●	●		●	
Apalachicola Dusky Salamander			●	●	●					
Northern Dusky Salamander							●	●	●	●
Spotted Dusky Salamander	●	●	●	●		●	●	●	●	●
Santeetlah Dusky Salamander							●		●	
Virginia Dusky Salamander								●		
Seal Salamander			●	●		●	●	●	●	●
Cumberland Dusky Salamander									●	
Carolina Mountain Dusky							●		●	
Allegheny Mountain Dusky								●	●	●
Blue Ridge Dusky Salamander							●	●	●	
Northern Two-lined Salamander								●		
Southern Two-lined Salamander	●	●	●	●	●	●	●	●	●	●
Blue Ridge Two-lined Salamander				●		●	●	●	●	
Three-lined Salamander	●	●	●	●	●		●	●	●	●
Long-tailed Salamander			●	●				●	●	●
Junaluska Salamander							●		●	
Dwarf Salamander	●	●	●	●	●					
Chamberlain's Dwarf Salamander		●		●		●				
Cave Salamander				●			●	●	●	●
Spring Salamander		●		●		●	●	●	●	●
Mud Salamander	●	●	●	●	●	●	●	●	●	●
Red Salamander	●	●	●	●		●	●	●	●	●
Patch-nosed Salamander				●		●				

COMMON NAME	LA	MS	AL	GA	FL	SC	NC	VA	TN	KY
SEASONAL WETLAND SALAMANDERS										
Frosted Flatwoods Salamander				●	●	●				
Reticulated Flatwoods Salamander			●		●					
Jefferson Salamander								●		●
Mole Salamander	●	●	●	●	●	●	●	●	●	●
Small-mouthed Salamander	●	●	●						●	●
Tiger Salamander	●	●	●	●	●	●	●	●	●	●
Mabee's Salamander						●	●	●		
Marbled Salamander	●	●	●	●	●	●	●	●	●	●
Spotted Salamander	●	●	●	●	●	●	●	●	●	●
Four-toed Salamander	●	●	●	●	●	●	●	●	●	●
Eastern Newt	●	●	●	●	●	●	●	●	●	●
Striped Newt				●	●					
FOREST TERRESTRIAL SALAMANDERS										
Green Salamander		●	●	●		●	●	●	●	●
Pygmy Salamander							●	●	●	
Red Hills Salamander			●							
Peaks of Otter Salamander								●		
Shenandoah Salamander								●		
Yonahlossee Salamander							●	●	●	
Blue Ridge Gray-cheeked							●			
South Mountain Gray-cheeked							●			
Southern Gray-cheeked						●	●			
Northern Gray-cheeked							●		●	
Eastern Red-backed Salamander							●	●	●	●
Big Levels Salamander								●		
Northern Slimy Salamander			●	●			●	●	●	
Chattahoochee Slimy Salamander				●						
Atlantic Coast Slimy Salamander				●		●	●	●		
White-spotted Slimy Salamander						●	●	●	●	
Southeastern Slimy Salamander			●	●	●					
Louisiana Slimy Salamander	●									
Mississippi Slimy Salamander	●	●	●						●	●
Ocmulgee Slimy Salamander				●						
Savannah Slimy Salamander				●						
South Carolina Slimy Salamander				●		●				
Valley and Ridge Salamander								●		
Shenandoah Mountain								●		
Cumberland Plateau Salamander								●	●	●
Southern Red-backed Salamander	●		●	●			●			
Wehrle's Salamander							●	●	●	●
Weller's Salamander							●	●	●	
Bay Springs Salamander*		●								
Tellico Salamander							●		●	
Pigeon Mountain Salamander				●						
Southern Appalachian Salamander				●		●	●		●	
Southern Ravine Salamander							●		●	●
Northern Ravine Salamander										●
Webster's Salamander	●	●	●	●		●				
Cheoah Bald Salamander							●			
Northern Zigzag Salamander		●	●						●	●
Southern Zigzag Salamander		●	●	●			●	●		
Red-legged Salamander				●			●		●	
Jordan's Red-cheeked Salamander							●		●	
Cow Knob Salamander								●		
TOTAL	23	27	42	53	26	39	60	52	56	34

* extinct species

Glossary

Adelgid An exotic insect (Hemiptera: Adelgidae: *Adelges tsugae*) introduced from Asia that has affected salamander populations in the Appalachian Mountains by killing hemlocks and Fraser firs.

Aestivation A period of inactivity during dry and/or hot periods in which animals wait for conditions suitable for feeding and other activities to return.

Albino An animal completely lacking the pigment that provides the color to skin and eyes. Animals lacking only dark pigment, or melanin, are often referred to as albinos but are more properly said to be amelanistic.

Anterior Referring to the end of the animal toward the head.

Anurans Amphibians that are frogs or toads.

Autotomy (*v.* autotomize) The breaking off of an appendage, such as the tail, as a means of escaping predation.

Biodiversity The number, distribution, and abundance of species within a given area.

Bioindicator A species whose health or condition, either at the individual level or at the population level, indicates the condition of the habitat or ecosystem as a whole.

Biomass The weight of living things in the environment; or the weight of all the animals in a population.

Biomonitor To measure and record the number, types, and characteristics of organisms.

Canthus rostralis A dark-bordered white line that runs from the eye to the nostril of the spring salamander (*Gyrinophilus porphyriticus*).

Caudates Amphibians that are salamanders.

Chytrid A fungus that many herpetologists consider responsible for the recent decline—and in some instances the extinction—of frog and toad populations, especially in high-elevation tropical habitats.

Chytridiomycosis The disease caused by the chytrid fungus that is sometimes fatal to amphibians.

Cirri (*sing.* cirrus) A pair of fleshy projections that point downward from the upper lip of some salamanders in the genus *Eurycea*. Cirri are most prevalent in males during the breeding season but are present in females of some species.

Cloaca (*adj.* cloacal) A single opening through which the urinary, digestive, and reproductive tracts exit the body.

Clutch A group of eggs laid together at one time by a single individual.

Cold-blooded A nontechnical term that refers to animals whose body temperature is largely determined by environmental conditions and the thermoregulatory behavior of the animal (*see* Ectotherm).

Courtship The series of behaviors that precedes mating.

Desmogs A colloquial term for dusky salamanders belonging to the genus *Desmognathus*.

Dorsolateral Referring to the area of the body between the back and sides.

Ecology The study of how organisms interact with their environment and other organisms.

Ectotherm (*adj.* ectothermic) An animal whose body temperature is largely determined by environmental conditions and the thermoregulatory behavior of the animal (*see* Cold-blooded).

Elastomer Material injected under the skin of amphibians as a colored liquid that hardens and becomes a visible marker that can be used to identify individuals.

Electromorphs Species that have been described as distinct from closely related species that are similar in appearance on the basis of genetic differences.

Endangered Referring to a species or population that is considered at risk of becoming extinct.

Endotherm An animal that maintains a constant body temperature primarily through the use of heat generated by its own metabolism (*see* Warm-blooded).

Ephemeral wetland Seasonal wetland habitats that usually hold water for only certain periods of the year.

Extinct Referring to species with no living individuals.

Fall Line The line separating the Piedmont from the Coastal Plain in the southeastern United States.

Family A taxonomic group containing two or more closely related genera.

Generalist An animal that does not specialize on any particular type of prey or is not restricted to a particular habitat.

Genus (*pl.* genera) A taxonomic grouping of one or more closely related species.

Herpetofauna The amphibians and reptiles that inhabit a given area.

Herpetologist A scientist who studies amphibians and reptiles.

Herpetology The scientific study of amphibians and reptiles.

Hibernaculum The microhabitat to which a salamander retreats during cold weather.

Hibernation A period of inactivity during cold periods; also known as "brumation" when referring to amphibians and reptiles.

Hybrid The offspring of mating between two different species.

Internal nares Openings in the roof of the mouth of salamanders that lead to the nasal passage.

Intergrade An intermediate form of a species resulting from mating and genetic mixing between individuals of two or more subspecies within a zone where their ranges overlap. Intergrade individuals may possess traits of two or more subspecies.

Mental gland A disk- to U-shaped gland on the underside of the chin of some plethodontid salamander males that produces pheromones used during courtship to entice females to mate.

Metamorphosis The process by which a larval aquatic salamander changes into a terrestrial juvenile.

Metamorph An amphibian that has recently undergone metamorphosis.

Microhabitat A small, defined area within the general habitat.

Nymph An aquatic insect larva.

Neoteny (*adj.* neotenic) A condition in which an adult salamander has certain traits such as gills that are typically found in aquatic larvae; *also* paedomorphosis.

Nocturnal Active at night.

Orthopteran A member of the insect order Orthoptera, which includes the katy-dids, grasshoppers, and crickets.

Oxbow A U-shaped lake that was once the main channel of a nearby river.

Paedomorphosis *See* Neoteny.

PARC Acronym for Partners in Amphibian and Reptile Conservation, the largest partnership group dedicated to the conservation of all amphibians and reptiles and their habitats.

Phylogeny The evolutionary and genealogical relationships among different groups and species of animals.

Physiographic region An area defined by the geological structure and general to-pography of the landscape.

PIT tag Acronym for "passive integrated transponder tag," a glass-encapsulated electronic device injected into the body cavity of animals for identification pur-poses.

Plethodontid (*adj.*) Referring to the family Plethodontidae.

Posterior Away from the head of an animal and toward the tail.

Radiotelemetry A method using a radiotransmitter attached to or implanted in an animal to track its movements by locating it using a directional antenna and radio receiver.

Reticulate A network pattern on the skin formed by an interconnecting series of light lines.

Rugose Creased or wrinkled.

Seep A permanently wet area created by groundwater slowly flowing out at the base of a steep bluff or between bedrock formations.

Silviculture Planned and managed growth and development of forests.

Specialist An animal restricted in its choice of diet or habitat.

Species Typically an identifiable and distinct group of organisms whose members breed and produce viable offspring under natural conditions.

Spermatophore A capsule of spermatozoa atop a gelatinous stalk deposited during courtship by the male of some species of salamanders.

Spring lizard A colloquial term for various species of southeastern mountain sala-manders, particularly the dusky salamanders, that are used for fish bait.

Subspecies A taxonomic unit, or "race," within a species, usually defined as mor-phologically distinct and occupying a geographic range that does not overlap with that of other races of the species. Subspecies may interbreed naturally in areas of geographic contact (*see* Intergrade).

Taxonomy The scientific field of classification and naming of organisms.

Troglobite A cave-dwelling animal.

Vaccination The process in which the male of some plethodontid salamander species uses his enlarged front teeth to scratch the skin of a female and then presses his mental gland against the abraded site to release a secretion that en-ters the female's body and presumably makes her receptive for mating.

Warm-blooded A nontechnical term referring to an animal that maintains its body temperature primarily through the use of metabolic heat (*see* Endotherm).

Woodland salamander A general term for all of the species in the genus *Plethodon*, which are terrestrial and have no aquatic larval stage.

Further Reading

Bailey, M. A., J. N. Holmes, K. A. Buhlmann, and J. C. Mitchell. 2006. *Habitat Management Guidelines for Amphibians and Reptiles of the Southeastern United States*. Montgomery, Ala.: Partners in Amphibian and Reptile Conservation. Tech. Publ. HMG-2.

Barbour, R. W. 1971. *Amphibians and Reptiles of Kentucky*. Lexington: University of Kentucky Press.

Bartlett, R. D., and P. P. Bartlett. 1999. *A Field Guide to Florida Reptiles and Amphibians*. Houston: Gulf Publishing Company.

———. 2006. *Guide and Reference to the Amphibians of Eastern and Central North America (North of Mexico)*. Gainesville: University of Florida Press.

Bean, J., A. Braswell, J. C. Mitchell, J. Harrison, and W. Palmer. 2010. *Amphibians and Reptiles of the Carolinas and Virginia*. 2nd ed. Chapel Hill: University of North Carolina Press.

Carmichael, P., and W. Williams. 2001. *Florida's Fabulous Reptiles and Amphibians*. Tampa, Fla.: World Publications.

Conant, R., and J. T. Collins. 1991. *A Field Guide to Reptiles and Amphibians of Eastern and Central North America*. 3rd ed. Boston: Houghton Mifflin.

Dodd, C. K. Jr. 2004. *The Amphibians of the Great Smoky Mountains National Park*. Knoxville: University of Tennessee Press.

Duellman, W. E., and L. Trueb. 1994. *Biology of Amphibians*. Baltimore: Johns Hopkins University Press.

Dundee, H. A., and D. A. Rossman. 1996. *Amphibians and Reptiles of Louisiana*. Baton Rouge: Louisiana State University Press.

Gibbons, J. W., and R. D. Semlitsch. 1991. *Guide to Reptiles and Amphibians of the Savannah River Site*. Athens: University of Georgia Press.

Jensen, J. B., C. D. Camp, J. W. Gibbons, and M. J. Elliott, eds. 2008. *Amphibians and Reptiles of Georgia*. Athens: University of Georgia Press.

Lannoo, M., ed. 2005. *Amphibian Declines: The Conservation Status of United States Species*. Berkeley: University of California Press.

Mitchell, J. C., A. R. Breisch, and K. A. Buhlmann. 2006. *Habitat Management Guidelines for Amphibians and Reptiles of the Northeastern United States*. Montgomery, Ala.: Partners in Amphibian and Reptile Conservation. Tech. Publ. HMG-2.

Mount, R. M. 1975. *The Reptiles and Amphibians of Alabama*. Tuscaloosa: University of Alabama Press.

Petranka, J. W. 1998. *Salamanders of the United States and Canada*. Washington, D.C.: Smithsonian Institution Press.

Semlitsch, R. D., ed. 2003. *Amphibian Conservation*. Washington, D.C.: Smithsonian Institution Press.

Stebbins, R. C., and N. W. Cohen. 1995. *A Natural History of Amphibians*. Princeton: Princeton University Press.

Zim, H. S., and H. M. Smith. 2001. *Reptiles and Amphibians. A Golden Guide*. New York: St. Martin's Press.

Acknowledgments

We are very grateful to the numerous herpetologists and others who provided information and editorial comments, including the following: Mark Bailey, Dick Bartlett, David Beamer, Jeff Beane, Steve Bennett, Jeff Boundy, Alvin Braswell, Dick Bruce, Kurt Buhlmann, Carlos Camp, Erica Crespi, Ken Dodd, Anne R. Gibbons, Gabrielle Graeter, Kristin Grayson, Judy Greene, Susan Lane Harris, Julian Harrison, Jeff Humphries, Bob Jones, Trip Lamb, Tom Luhring, John MacGregor, Bruce Means, Brian Miller, Paul Moler, Matt Niemiller, Tom Pauley, Leslie Rissler, Steve Roble, Travis Ryan, David Scott, Ray Semlitsch, Dirk Stevenson, Brian Todd, Jayme Waldron, Margaret Wead, Patricia West, J. D. Willson, Addison Wynn. Mindy Conner did an excellent job of copyediting this book and others in the series. John Jensen read an early draft of the entire manuscript and provided excellent suggestions for improvement. Margaret Wead scanned slides and organized digital images for the book and was helpful in a variety of other ways with the organization and preparation of files. Susan Lane Harris created the indexes and provided several editorial comments that improved the manuscript.

Herpetologists throughout the country provided a spectacular array of salamander slides and images. The excellent color images of species and habitats generously offered by so many experts in animal photography have added tremendously to the value of this book. We thank the following for providing images and slides: Mark Bailey, Jaime Barichivich, Brady Barr, David Beamer, Jeff Beane, Jeff Boundy, Alvin Braswell, E. D. Brodie Jr., Will Brown, Alan Cameron, Carlos Camp, E. D. Corey, Alan Cressler, David Dennis, Ken Dodd, Sarah Eddy, Kevin Enge, Don Forester, Mike Gibbons, Whit Gibbons, Gabrielle Graeter, Mike Graziano, Jeff Hall, Julian Harrison, Jesse Helton, Aubrey M. Heupel, E. Pierson Hill, Bill Hopkins, Jeff Humphries, Vladimir Ischenko, John Jensen, Robin Jung Brown, Trip Lamb, Tom Luhring, John MacGregor, John Maerz, Mike Marchand, Johnathon Mays, Kathryn Pawlik McCoard, Bruce Means, Brian Miller, Joe Mitchell, Paul Moler, Brad Moon, Matthew Niemiller, Justin Oguni, Bill Peterman, James Petranka, Lisa Powers, Todd Pusser, Steve Roble, Francis Rose, Kevin Saunders, David Scott, Ray Semlitsch, Dirk Stevenson, Steve Tilley, Brian Todd, Josef C. Uyeda, Robert Wayne Van Devender, John White, Lori A. Williams, J. D. Willson, Robert Zappalorti. Special thanks goes to DigiMorph Co., A National Science Foundation Digital Library at the University of Texas at Austin, for allowing us to use the image of the tiger salamander skeleton.

Joe Mitchell especially thanks Susan C. Walls for her encouragement and support during the writing of this book and regrets that she was unable to join us on this project. He thanks his children—Tanya, Joshua, Justin, and Lisa—for their support over the years by letting their old man venture around the Southeast in search of amphibians and reptiles.

Whit Gibbons is especially appreciative of the support and understanding of his family during the preparation of the book: Carolyn, Laura, Michael, Jennifer, Allison, Parker, and Gilbey Gibbons; Susan Lane, Keith, and Nicholas Harris; Jennifer, Jim, and Samuel High. Finally, because the book supports the efforts of Partners in Amphibian and Reptile Conservation (PARC) to promote education about reptiles and amphibians, we thank the many PARC members who offered encouragement, advice, and enthusiastic support.

Credits

The authors would like to thank the following individuals and organizations for providing photographs:

Brady Barr
Photograph on page 21 (top).

David Beamer
Photographs on pages 133 (middle) and 228.

E. D. Brodie Jr.
Photographs on pages 20 (top left and bottom right), 66, 71 (top right), 189 (top), and 254.

Alan Cameron
Photograph on page 221 (bottom).

Carlos Camp
Photographs on pages 3 (bottom), 117 (bottom), and 129 (bottom left).

E. D. Corey
Photographs on pages 72, 134 (top), and 239 (top right).

Alan Cressler
Photographs on pages viii, 30, 31, 32, 35 (right), 38 (top), 93 (left), 131, 159 (top), and 236.

David Dennis
Photographs on pages 1 (right), 9 (right), 10 (left), 11 (right), 13 (top), 15 (both), 16 (bottom right), 18 (top), 19, 20 (middle and bottom left), 22 (middle), 26, 48 (bottom right), 51, 55, 59, 61, 65 (top), 71 (bottom), 89, 90 (top), 93 (right), 97 (top), 101 (top), 104 (bottom), 111 (left), 114 (left), 117 (top), 123 (top), 125 (middle), 129 (top), 132, 134 (bottom), 136, 137 (top), 141 (both), 147 (bottom), 150, 151, 152, 153 (top and bottom left), 157 (bottom), 158, 161 (left), 164, 168 (left), 175 (top), 179 (middle and bottom), 180 (top), 181, 187 (top), 199 (right), 203, 205

(bottom right), 223, 224, 227 (bottom), 246 (first, third, and fourth in left column), 266, 280, 285, 286, 288, and 294 (bottom).

DigiMorph Co., University of Texas, Austin
Photograph on page 6.

Ken Dodd
Photographs on pages 53 (left), 140, 143 (bottom), 176, and 214 (bottom).

Sarah Eddy
Photograph on page 10 (right).

Don Forester
Photographs on pages 101 (bottom) and 149 (bottom).

Mike Gibbons
Photograph on page 300.

Gabrielle Graeter
Photographs on pages 7 (top left) and 54 (top).

Mike Graziano
Photographs on pages 7 (bottom left), 40, 41, 60 (top and bottom), 70, 78 (top), 100, 107, 109, 138, 147 (top), 170, 178, 179 (top), 184 (bottom), 188, 190 (bottom), 195 (bottom), 205 (top), 208 (right), 209 (top), 214 (top), 215, 231 (top), 244, 246 (bottom of left column), 258, 276, and 291.

Jeff Hall
Photographs on pages 62, 68, 155 (top), 166, 202, and 295 (top).

Jesse Helton
Photograph on page 162 (bottom right).

Aubrey M. Heupel
Photographs on pages 3 (top) and 124.

E. Pierson Hill
Photographs on pages 47 (top) and 81 (both).

Bill Hopkins
Photograph on page 293.

Vladimir Ischenko
Photograph on page 18 (bottom).

John Jensen
Photographs on pages 33 (right), 35 (left), 57, 63 (top), 67 (top), 113 (top), 114 (right), 126 (right), 146 (right), 171, 216, 239 (top left), 246 (second in left column), 270, and 274.

Robin Jung Brown
Photograph on page 304.

Trip Lamb
Photographs on pages i, ii, iii, v, vi, vii, 4, 12, 44, 95, 103, 115, 122, 125 (bottom), 139, 155 (middle and bottom), 173, 196, 213 (bottom), 217, 235, 238, 241, 245, 248, and 298 (top).

John MacGregor
Photographs on pages 13 (bottom), 17, 22 (top), 42, 50 (bottom), 73, 85, 87, 99, 108, 111 (right), 121, 125 (top), 130 (top), 135, 177 (bottom), 180 (bottom), 183 (both), 186 (bottom), 189 (bottom), 200 (top), 206 (top), 207, 211 (second from bottom), 225, 226, 249 (top), 251, 256, 261, 263, 271, 272, 278, 282, 283 (right), 284 (top), 287 (top), and 294 (top).

John Maerz
Photograph on page 61.

Mike Marchand
Photograph on page 301 (right).

Johnathon Mays
Photographs on pages 7 (top right), 118, 120 (right), 198, and 201 (middle).

Kathryn Pawlik McCoard
Photographs on pages 128 (top) and 140.

Bruce Means
Photograph on page 227 (top).

Brian Miller
Photograph on page 284 (bottom).

Joe Mitchell
Photographs on pages 34 (left), 38 (bottom), 145, 149 (top), 204 (top), 249 (bottom right), 296 (top left, bottom left, and bottom right), 301 (left), 302, 303, and 305.

Brad Moon
Photographs on pages 11 (left), 16 (top left, middle, and bottom left), 37 (middle), 48 (bottom left), 79 (bottom), 80 (top), 297, and 298 (bottom).

Matt Niemiller
Photographs on pages 29, 79 (top), 91 (top), 130 (bottom), 162 (bottom left), 168 (right), 169 (bottom), and 219.

Justin Oguni
Photographs on pages 159 (bottom) and 222.

Bill Peterman
Photographs on pages 23 (bottom), 28 (top), 52, 157 (top), 162 (top), 221 (top), 249 (bottom left), 260, 283 (left), and 287 (bottom).

James Petranka
Photographs on pages 27 (right), 142, and 213 (top).

Lisa Powers
Photographs on pages 1 (left), 14, 33 (top left), 39, 98, 167, and 191 (top).

Todd Pusser
Photograph on page 47 (bottom).

Steve Roble
Photograph on page 17.

Francis Rose
Photograph on page 20 (top right).

Kevin Saunders
Photographs on pages 9 (left) and 257 (top).

David Scott
Photographs on pages 211 (bottom), 246 (bottom right), and 292.

Dirk Stevenson
Photographs on pages 37 (bottom), 75 (top), 104 (top), 185, 187 (bottom), 192 (top), 201 (top), 212, 295 (bottom), and 296 (top right).

Steve Tilley
Photographs on pages 23 (bottom), 127 (both), 133 (top), 237, and 273.

Brian Todd
Photographs on pages 5, 24, 25, 102, 119 (top), and 144.

Josef C. Uyeda
Photographs on pages 58 (bottom), 116 (both), 120 (left), and 143 (top).

Robert Wayne Van Devender
Photographs on pages 7 (bottom right), 27 (left), 28 (bottom), 48 (top), 50 (top), 54 (bottom), 56 (both), 63 (bottom), 64 (both), 65 (middle and bottom), 67 (bottom), 69, 74, 75 (middle left, middle right, and bottom), 88 (both), 90, 91 (bottom), 92, 105 (both), 106, 110, 119 (bottom), 123 (bottom), 126 (left), 133 (bottom), 137 (bottom), 160, 161 (right), 163, 175 (bottom), 177 (top), 184 (top), 191 (bottom), 193, 195 (top), 201 (bottom), 210 (top), 211 (second and third from top), 229, 232, 234, 239 (bottom left and bottom right), 243 (both), 246 (fifth and sixth in left column and middle), 251, 253, 255, 257 (bottom), 259, 265, 268, 269, 275, 277 (both), 281, 289, and 290.

John White
Photographs on pages 165 (left), 186 (top), 190 (top), 205 (bottom left), 211 (top), 231 (bottom), and 252.

Lori A. Williams
Photographs on pages 34 (right), 53 (right), and 200 (bottom).

J. D. Willson
Photographs on pages 7 (middle), 58 (top), 82, 83, 113 (bottom), 146 (left), 153 (bottom right), 156, 165 (right), 169 (top), 182, 192 (bottom), 197, 199 (left), 206 (bottom), 208 (left, 210 (bottom), 230, and 240.

Robert Zappalorti
Photographs on pages 71 (top left), 194, 204 (bottom), and 209 (bottom).

Index of Scientific Names

Boldface page numbers refer to species accounts. *Italicized* page numbers refer to illustrations.

Ambystoma barbouri, **97–100**
Ambystoma bishopi, **175–178**
Ambystoma cingulatum, **175–178**
Ambystoma jeffersonianum, **179–182**
Ambystoma mabeei, **195–197**
Ambystoma maculatum, **201–204**
Ambystoma malvortium, 194
Ambystoma mexicanum, 20
Ambystoma opacum, **198–200**, 297
Ambystoma talpoideum, **183–186**, 297
Ambystoma texanum, **187–190**
Ambystoma tigrinum, **191–194**, 298
Amphiuma means, **47–51**
Amphiuma pholeter, **47–51**
Amphiuma tridactylum, **47–51**
Andrias davidianus, 21
Andrias japonicas, 21
Aneides aeneus, **219–221**

Bolitoglossa mexicana, 20

Cryptobranchus alleganiensis, **52–55**
Cryptobranchus alleganiensis alleganiensis, 55

Desmognathus abditus, **132–136**
Desmognathus aeneus, **101–103**
Desmognathus apalachicolae, **121–123**
Desmognathus auriculatus, **104–106**
Desmognathus carolinensis, **132–136**
Desmognathus conanti, **124–128**
Desmognathus folkertsi, **113–116**
Desmognathus fuscus, **124–128**
Desmognathus imitator, **107–109**, 294

Desmognathus marmoratus, **56–59**
Desmognathus monticola, **129–131**
Desmognathus ochrophaeus, **132–136**
Desmognathus ocoee, **117–120**
Desmognathus orestes, **132–136**
Desmognathus planiceps, **124–128**
Desmognathus quadramaculatus, **113–116**
Desmognathus santeetlah, **124–128**
Desmognathus welteri, **110–112**
Desmognathus wrighti, 21, **222–224**

Eurycea aquatica, **140**
Eurycea bislineata, **137–140**
Eurycea chamberlaini, **152–156**
Eurycea cirrigera, **137–140**
Eurycea guttolineata, **144–146**, 149
Eurycea junaluska, **150–151**
Eurycea longicauda, 146, **147–149**
Eurycea longicauda longicauda, 149
Eurycea longicauda melanopleura, 149
Eurycea lucifuga, 90, **157–159**
Eurycea quadridigitata, **152–156**
Eurycea wilderae, 140, **141–143**

Gyrinophilus gulolineatus, **90–93**
Gyrinophilus palleucus, **90–93**
Gyrinophilus palleucus necturoides, 93
Gyrinophilus palleucus palleucus, 93
Gyrinophilus porphyriticus, **160–163**
Gyrinophilus porphyriticus danielsi, 161, 163
Gyrinophilus porphyriticus dunni, 161, 163
Gyrinophilus porphyriticus duryi, 161, 163

Gyrinophilus porphyriticus porphyriticus, 161, 163

Haideotriton wallacei, **87–89**
Hemidactylium scutatum, **205–208**

Necturus alabamensis, **63–66**
Necturus beyeri, **63–66**
Necturus lewisi, **63–66**
Necturus maculosus, **70–73**, 293
Necturus maculosus louisianensis, 70, 72, 73
Necturus maculosus maculosus, 73
Necturus punctatus, **67–69**
Notophthalmus perstriatus, **214–216**
Notophthalmus viridescens, **209–213**
Notophthalmus viridescens dorsalis, 209–213
Notophthalmus viridescens piaropicola, 209–213
Notophthalmus viridescens viridescens, 209–213

Oedipina poelzi, 20

Phaeognathus hubrichti, **225–227**
Plethodon ainsworthi, 4, **267**
Plethodon amplus, **237–240**
Plethodon angusticlavius, **284**
Plethodon aureolus, **268–269**
Plethodon chattahoochee, **245–251**
Plethodon cheoah, **280–281**
Plethodon chlorobryonis, **245–251**
Plethodon cinereus, **241–244**, 294
Plethodon cylindraceus, **245–251**
Plethodon dorsalis, **282–284**
Plethodon electromorphus, **275–277**
Plethodon glutinosus, **245–251**
Plethodon grobmani, **245–251**
Plethodon hoffmani, **252–254**

Plethodon hubrichti, 228–230
Plethodon jordani, 237, 274, 286, 287–288, 294
Plethodon jordani teyahalee, 274
Plethodon kentucki, 255–257
Plethodon kisatchie, 245–251
Plethodon meridianus, 237–240
Plethodon metcalfi, 237–240
Plethodon mississippi, 245–251
Plethodon montanus, 237–240
Plethodon ocmulgee, 245–251
Plethodon oconaluftee. See *Plethodon teyahalee*
Plethodon petraeus, 270–272
Plethodon punctatus, 289–291
Plethodon richmondi, 275–77
Plethodon savannah, 245–251
Plethodon serratus, 258–260
Plethodon shenandoah, 231–233, 294
Plethodon sherando, 241–244
Plethodon shermani, 285–286
Plethodon teyahalee, 273–274
Plethodon variolatus, 245–251

Plethodon ventralis, 282–284
Plethodon virginia, 252–254
Plethodon websteri, 278–279
Plethodon wehrlei, 261–263
Plethodon welleri, 264–266
Plethodon yonahlossee, 234–236
Pseudobranchus axanthus, 74–77
Pseudobranchus axanthus axanthus, 74, 77
Pseudobranchus axanthus belli, 74, 77
Pseudobranchus striatus, 74–77
Pseudobranchus striatus lustricolus, 74, 77
Pseudobranchus striatus spheniscus, 74, 77
Pseudobranchus striatus striatus, 77
Pseudotriton montanus, 164–166
Pseudotriton montanus diastictus, 164–166
Pseudotriton montanus flavissimus, 164–166
Pseudotriton montanus floridanus, 164–166

Pseudotriton montanus montanus, 164–166
Pseudotriton ruber, 167–170
Pseudotriton ruber nitidus, 167–170
Pseudotriton ruber ruber, 167–170
Pseudotriton ruber schencki, 167–170
Pseudotriton ruber vioscai, 167–170

Ranodon sibericus, 21

Salamandra salamandra, 18, 21
Salamandrella keyserlingii, 18–19
Siren intermedia, 78–80
Siren intermedia intermedia, 78–80
Siren intermedia nettingi, 78–80
Siren lacertina, 81–83
Stereochilus marginatus, 60–62

Triturus marmoratus, 19
Typhlotriton spaeleus, 19, 20, 21

Urspelerpes brucei, 21, 171–172

Index of Common Names

Boldface page numbers refer to species accounts. *Italicized* page numbers refer to illustrations.

alligator salamander. *See* hellbender

amphiuma, 6, **47–51**, 77, *295*
    and biting, 8, 11
    geographic range of, 21
    habitat of, *32*, 34
    identification of, 21, 25, 28, 29, 76, 78, 82
    longevity of, 17
    one-toed, 34, **47–51**
    as predator, 9
    as prey, 11
    and reproduction, 14
    sounds made by, 12
    three-toed, **47–51**
    two-toed, **47–51**

Appalachian salamander, southern, 238, 248, **273–274**
    and hybridization, 281, 285, 287

Bay Springs salamander, 4, **267**

Berry Cave salamander, 39, **90–93**

Big Levels salamander, **241–244**

black-bellied salamander, 57, 59, **113–116**
    dwarf, **113–116**
    habitat of, 35
    identification of, 125, 130
    as predator, 119, 128, 140, 143, 288

Black Mountain salamander, *13*, **110–112**, 130

blue-spotted salamander, *182*

brown-backed salamander, 140

cave salamander, 90, 148, **157–159**, 251

Cheoah Bald salamander, **280–281**

Chinese giant salamander, 19, 21

Congo eel. *See* amphiuma

Cow Knob salamander, 248, **289–291**

Cumberland Plateau salamander, 39, 238, 248, **255–257**

Desmognathus, 7
    identification of, *28*, 56, 57, 114, 225

devil dog. *See* hellbender

dusky salamander
    Apalachicola, **121–123**
    common, **124–128**, 130
    habitat of, 35, 36
    identification of, 25, 26, 28, 110, 139, 142, 150, 160, 222, 225
    northern, 17, 28, 105, 106, **124–128**, 134
    as prey, 72, 162–163
    Santeetlah, **124–128**
    southern, 30, **104–106**, 121
    spotted, 105, 118, 121–122, **124–128**
    Virginia, 28, **124–128**

dwarf black-bellied salamander, **113–116**

dwarf salamander, 1, **152–156**
    Chamberlain's, **152–156**
    habitat of, 34, 35
    identification of, 24–27, 60–61, 171, 206

dwarf siren, **74–77**
    identification of, 25, 68, 78, 82
    northern, **74–77**
    southern, **74–77**

European fire salamander, *18*, 21

European olm, 21

fire-bellied newt, 298

flatwoods salamander, 36, **175–178**, 297
    frosted, *41*, **175–178**
    reticulated, **175–178**

four-toed salamander, 25, 34, 152, **205–208**

Georgia blind salamander, *31*, 39, **87–89**

grampus. *See* hellbender

gray-cheeked salamander, 23, **237–240**
    Blue Ridge, *27*, **237–240**
    northern, 23, **237–240**, 261, 275
    southern, 23, **237–240**
    South Mountain, **237–240**

greater siren, *3*, 48, 76, 78, **81–83**

green salamander, *2*, 36, **219–221**

grotto salamander, 19, 20, 21, 87, 93

hellbender, 6, **52–55**, *293*, *295*
    and biting, 8
    eastern, **55**
    geographic range of, 21
    habitat of, 34
    identification of, 21, 29, 70
    longevity of, 17
    Ozark, **55**
    as predator, 9, 10
    and reproduction, 14–15

imitator salamander, **107–109**

Japanese giant salamander, 21
Jefferson salamander, **179–182**,
208
identification of, 98, 192,
198, 202
Jordan's red-cheeked salamander,
9, 109, **287–288**
former subspecies of, 237,
240, 274, 281, 286
and hybridization, 269, 273
identification of, 280
and reproduction, 15, 274
Junaluska salamander, 35, **150–151**

lamp eel. *See* amphiuma
lesser siren, **78–80**
eastern, **78–80**
identification of, 48, 76,
81–82
western, **78–80**
long-tailed salamander, 39, 146,
147–149
identification of, 138, 142,
144, 157

Mabee's salamander, 7, **195–197**
many-lined salamander, **60–62**
marbled newt, *19*
marbled salamander, 7, **198–200**,
293, 295, 297
habitat of, 33
identification of, 27, 98, 176,
180, 184, 188, 192, 196,
202
as predator, 11–12, 204, 208
as prey, 11–12
and reproduction, 14, 29
Mexican axolotyl, *20*
Mexican mushroom salamander, *20*
mole salamander, 17, **183–186**, 297
identification of, 25, 98, 176,
196, 198, 202
mole salamanders (Ambystomatidae), 6, 7, 10, 12, 14
habitat of, *32,* 33–36, 37, 303

identification of, 27, 28, 64,
175, 179, 187, 188, 191,
195
mollyhugger. *See* hellbender
mountain dusky salamander, 23,
107, 120, **132–136**
Allegheny, **132–136**
Blue Ridge, **132–136**
Carolina, **132–136**
Cumberland, **132–136**
identification of, 125, 234
mud devil. *See* hellbender
mudpuppy, 7, 8, **70–73**, 293. *See
also* waterdog
common, **73**
geographic range of, 21
habitat of, 32, 34
identification of, 25, 26, 28,
52, 64, 68
as pet, 299
Red River, **70,** 72
and reproduction, 14
mud salamander, *12,* 61, **164–166**
eastern, **164–166**
Gulf Coast, **164–166**
habitat of, 34, 35
identification of, 27, 160,
168, 206
midland, **164–166**
rusty, **164–166**

newt, 8, 21, 29, 208, 295. *See also*
newt, eastern; striped newt
defense mechanisms of, 11,
119, 136
habitat of, *32, 33,* 34–36
identification of, 26–28
and reproduction, 15, 17
newt, eastern, 11, **209–213**
broken-striped, **209–213,**
215
central, **209–213**
peninsula, **209–213**
red-spotted, 7, 8, 12, 17, *34,*
204, **209–213**

Ocoee salamander, 26, **117–120**,
162
identification of, 102, 107,
134, 222
ornate chorus frogs, 5

patch-nosed salamander, 21,
171–172
Peaks of Otter salamander,
228–230
permanent cave salamanders,
90–93
Pigeon Mountain salamander, 39,
270–272
Plethodon, 7, 8–9, 274
defense mechanisms of, 12,
236
habitats of, 36, 303
identification of, 29, 134
as prey, 162–163
and reproduction, 42, 240,
256, 263
territories of, 250, 254, 260,
262
territory marking by, 12–13
Plethodontidae, 18, 27
decline of, 303
defense mechanisms of, 12
diversity of, 6, 8
as "electromorphs," 26
geographic distribution of,
18, 19
habitats of, 31, 35, *39*
and hybridization, 273
identification of, 9, 27, 262
as predator, 10
and reproduction, 15, 61
sounds made by, 12
pygmy salamander, 21, **222–224**
habitat of, 39, 303
identification of, 27, 102, 264

ravine salamander, 252, **275–277**
northern, **275–277**
southern, 241–242, **275–277**

red-backed salamander, 9, 233,
      241–244
   Big Levels, **241–244**
   eastern, 22, 27, 238,
      241–244, 248, 252, 258,
      260, 264
   identification of, 228, 232,
      234
   southern, 27, 102, 222, 238,
      242, **258–260**, 282
red-cheeked salamander. *See* Jordan's red-cheeked salamander
Red Hills salamander, **225–227**
red-legged salamander, 39,
      285–286
red salamander, **167–170**
   black-chinned, **167–170**
   Blue Ridge, **167–170**
   defense mechanisms of, *11*,
      12, 166
   habitat of, 35, 36
   identification of, 27, 160,
      164–165, 206
   northern, **167–170**
   as predator, 123, 128, 143
   as prey, 162–163
   southern, **167–170**

seal salamander, 110, 119, 125,
      129–131, 134
seepage salamander, **101–103**, 222,
      224
Shenandoah Mountain salamander, 23, **252–254**
Shenandoah salamander, *22*,
      231–233
shovel-nosed salamander, **56–59**,
      114, 125, 130, 143
Siberian newt, 18–19
siren, 8, 77. *See also* greater siren;
   lesser siren
      geographic range of, 21
      habitat of, *32*, 34
      identification of, 21, 25,
         26, 28

as pet, 299
as predator, 69
as prey, 11
and reproduction, 14
sounds made by, 12
slimy salamander, *10*, 26, **245–251**,
   257, 286
   Atlantic Coast, **245–251**
   Chattahoochee, **245–251**
   defense mechanisms of, 11
   habitat of, 35
   identification of, 27, 232,
      238, 275
   Louisiana, **245–251**
   Mississippi, **245–251**
   northern, *17*, *22*, **245–251**,
      255, 261, 289
   Ocmulgee, **245–251**
   as prey, 158
   Savannah, **245–251**
   South Carolina, **245–251**
   southeastern, **245–251**
   white-spotted, 228, **245–251**,
      289
small-mouthed salamander,
      187–190
   identification of, 98, 180,
      184, 192, 202
snot otter. *See* hellbender
spotted salamander, 7, *16*,
      201–204, *304*
   decline of, 303
   habitat of, 33
   identification of, 98, 176,
      180, 184, 188, 196,
      198
   longevity of, 17
   as predator, 208
   as prey, 11–12
   and reproduction, *16*, 29
spring salamander, *10*, 28, 40,
      160–163
   Blue Ridge, **160–163**
   Carolina, **160–163**
   habitat of, 35

identification of, 28, 90,
      164–165, 168
Kentucky, **160–163**
northern, **160–163**
as predator, 103, 109, 112,
      116, 119, 128, 131, 136,
      143, 224, 240, 288
streamside salamander, **97–100**
   identification of, 180, 184,
      188, 192, 198
striped newt, 26–27, 36, 210,
      214–216, 303

Tellico salamander, 248, **268–269**
Tennessee cave salamander, 39,
      87, **90–93**, 160
   pinkish, **93**
   spotted Big Mouth, **93**
three-lined salamander, 35,
      144–146, 149
   identification of, 26–27, 138,
      142, 148
tiger salamander, 7, **191–194**, 294,
   297, *300*
   activity of, 8
   barred, **194**
   defense mechanisms of, 11
   habitat of, *33*, 36, 39
   as pet, 298
   as predator, 9, 189, 197
   as prey, 11–12
   and reproduction, 29
two-lined salamander, **137–140**, 151
   Blue Ridge, *17*, 138, 140,
      141–143, 150
   habitat of, 35, 36, 39
   identification of, 24, 26–27,
      134, 171, 206
   northern, **137–140**, 143
   as prey, 58, 72, 162–163, 166
   southern, 27, **137–140**, 142,
      152

Valley and Ridge salamander, 23,
      241–242, **252–254**

waterdog, 8, *29*, **63–66**
    Black Warrior River, **63–66**, 70
    dwarf, 64, **67–69**
    geographic range of, 21
    Gulf Coast, **63–66**
    habitat of, 34
    identification of, 25, 26, 68
    Neuse River, **63–66**, 68
    as pet, 299

Webster's salamander, 29, 258, **278–279**, 282, 284
Wehrle's salamander, 248, **261–263**, 275, 289
Weller's salamander, **264–266**
white-spotted salamander.
    See Cow Knob salamander
woodland salamanders.
    See *Plethodon*

Yonahlossee salamander, *4*, **234–236**, 238, 264

zigzag salamander, 39, 279, **282–284**
    identification of, 26–27, 278
    northern, 275, **282–284**
    Ozark, 284
    southern, 258, **282–284**